Plant bugs on fruit crops in Canada

Heteroptera: Miridae

Leonard A. Kelton
Biosystematics Research Institute
Ottawa, Ontario

Research Branch
Agriculture Canada

Monograph No. 24
1982

Canadian Cataloguing in Publication Data

Kelton, Leonard A.
Plant bugs on fruit crops in Canada

(Monograph ; no. 24)
Includes bibliographical references and index.

1. Miridae. 2. Fruit—Diseases and pests—Canada.
3. Insects, Injurious and beneficial—Canada.
I. Canada. Agriculture Canada. Research Branch.
II. Title. III. Series: Monograph (Canada.
Agriculture Canada) ; no. 24.

QL523.M5K4   595.7′54   C83-097200-5

Contents

Acknowledgments

I would like to express my thanks to the following colleagues from Agriculture Canada's Research Stations for making available their research records, specimens, and expertise: L. S. Thompson (Charlottetown, P.E.I.), A. W. MacPhee (Kentville, N.S.), G. W. Wood (Fredericton, N.B.), R. O. Paradis (Saint-Jean, Que.), D. R. Pree, (Vineland Station, Ont.), R. D. McMullen (Summerland, B.C.), and W. T. Cram (Vancouver, B.C.).

I gratefully acknowledge the cooperation of the Directors, Drs. L. B. MacLeod, Charlottetown; G. C. Russell, Summerland; former Directors, Drs. A. J. McGinnis, Vineland Station; J. R. Wright, Kentville; G. M. Weaver, Fredericton; and J.-J. Jasmin, Saint-Jean; also J. A. Archibald, Ontario Ministry of Agriculture and Food, Vineland Station, for providing research facilities while working at the Research Stations during the years 1974–1977.

I am grateful to Dr. R. C. Froeschner, U.S. National Museum, Washington, D.C., and T. J. Henry, SEL. USDA, U.S. National Museum, Washington, D.C., for making available the collections in their care.

Special thanks are due to D. Brown, my technician, for curating the specimens and preparing the distribution maps, and to L. Yusyk and S. Rigby of this Institute, for many illustrations of adult Miridae. I also thank Drs. J. F. McAlpine and C. M. Yoshimoto for critically reviewing the manuscript.

Introduction

Plant bugs on fruit crops are of great economic importance to man because they are either harmful or beneficial. The harmful bugs are those that suck out the plant juices and injure the plant or damage the fruit; these are the phytophagous species, and losses caused by them often amount to millions of dollars annually. The beneficial bugs are those that prey on and destroy the arthropods that feed on the plant or the fruit; these are the predaceous species, and without them damage to fruit crops would be considerably greater.

In economic terms the effect of the phytophagous species seem to far outweigh the beneficial effect of the predaceous species. Large populations of phytophagous species reduce plant vigor, transmit virus diseases, and cause fruit drop and malformed fruit. The resulting effect is reduced yield and lower grades of fruit. Often the bugs may completely destroy the fruit crop. To control these pests the fruit grower has two choices: a chemical spray program, which cuts into the returns from the fruit crop, or a natural control program relying on predators of various arthropods that feed on the fruit crop.

Because these predators prey on mites, aphids, psyllids, leafhoppers, small lepidopterous larvae, and other soft-bodied arthropods that feed on the plant or the fruit, they substantially reduce the damaging effect of the harmful pests. In sufficient numbers, the predators may reduce or eliminate the need of a chemical spray.

Plant bugs are not well known despite their economic importance, the large numbers found on the fruit crops, and the fact that they have been with us for a long time. Doubt exists whether they are harmful or beneficial, confusion exists in naming them, and little is known of their biology. The aim of this bulletin is twofold: 1) to help economic research scientists and fruit growers identify the plant bugs found on the fruit crops, and 2) to provide information on their habits and biology. Once these prerequisites are established, the fruit growers must decide, individually or in consultation with the research scientists, the type of program necessary to effectively control the pests.

In recent years special efforts were made to collect the plant bugs associated with the fruit crop from the fruit growing areas of Canada. The major fruit growing areas are the Okanagan Valley of British Columbia, the Niagara Peninsula of Ontario, the apple growing area of southwestern Quebec, and the Annapolis Valley of Nova Scotia. As a result of these collections, knowledge of the species associated with the fruit crops has been expanded and new information on their habits and biology has been obtained. For plant bugs collected on apple trees in southwestern Quebec see Braimah et al. (1982).

This faunal study shows that 81 species of Miridae from 34 genera have been collected on cultivated and native fruit crops in Canada. Of this number, 24 species are phytophagous, 47 are predaceous, and 10 are phytophagous and predaceous. The fruit crops investigated are apple, pear, peach, plum, apricot, sweet cherry, sour cherry, black cherry, chokecherry, pin cherry, mulberry, raspberry, thimbleberry, loganberry, blackberry, currant, gooseberry, blueberry, serviceberry, viburnum, elderberry, cranberry, strawberry, and grape. Most of the plant bugs collected are endemic to the Nearctic region but 15 species are accidental introductions from Europe.

The 81 species collected represent approximately 6% of the total number of Miridae species believed to occur in Canada. Adults of most species treated here are illustrated, and male claspers of closely related species are figured. Brief descriptions of adults, biology, and distributions are included. Keys to subfamilies, genera, and species are also provided.

Collecting, preserving, and identifying specimens

Three methods are used to collect Miridae on fruit plants. The sweeping method, using a regular sweep net, is used on young flexible branches of trees and shrubs, and on plants that grow close to the ground. Because mirids are fragile and delicate insects, sweeping must be done carefully to avoid damaging the bugs in the net. Leaves, fruit, and other debris, often picked up in sweeping, can damage the specimens in the net if sweeping is prolonged; therefore, the bugs should be picked out of the net frequently with an aspirator. Sweeping should be done under dry conditions, as moisture in the net will mat and ruin the specimens.

The beating stick and sheet method is used for collecting mirids on branches of mature fruit trees. It is the most productive method and collections from isolated branches give accurate host associations. For this method the sheet is held under a branch and the branch is sharply struck with the stick. The bugs are jarred loose, fall on the sheet, and are picked off the sheet quickly with an aspirator.

The third method is to search for individual specimens on the trunks, limbs, and other parts of fruit plants. The bugs that inhabit the bark are naturally well camouflaged and sit on the bark or hide in the crevices. When disturbed they move a short distance and if the collector is quick enough the bugs are picked up directly with the aspirator.

The collected specimens are killed promptly in cyanide and mounted. If they cannot be mounted at the end of the day, they may be stored for several weeks between layers of cellulose cotton in pill boxes. Each pill box is labeled with pertinent information about the specimens such as place collected, date collected, and host plant. Before mounting the stored specimens, the pill boxes containing the specimens are placed in a relaxing container and the bugs relaxed.

All mirids should be mounted on narrow triangular bristol board points. The tip of the point is bent to fit the angle of the thorax so that the specimen will be level when mounted. Only the tip of the point should be covered with glue and the point attached to the right side of the thorax above the middle coxa. Miridae should not be pinned through the body, and they should never be placed in alcohol. For additional details on collecting and preserving techniques of other insects, see Martin (1977).

Nearly all the plant bugs associated with fruit crops may be identified with the aid of a high-powered hand lens. In a few instances a binocular stereomicroscope will be required to examine the closely related species, and in particular viewing the male claspers. Specimens that can not be identified, or specimens that need confirmation, should be forwarded to: National Identification Service, Biosystematics Research Institute, Agriculture Canada, Ottawa, Ontario.

Most of the material for this book was collected by me and is in the Canadian National Collection of Insects, Ottawa. Some records were obtained from Agriculture Research Stations, Vancouver and Summerland, British Columbia; Vineland Station and Ottawa, Ontario; St. Jean, Quebec; Fredericton, New Brunswick; Kentville, Nova Scotia; and Charlottetown, Prince Edward Island. My field work usually commenced in the second week of June and was carried on throughout July, August, and early September. Collection localities for each species are indicated by dots and squares on the maps, and the general range is given under the distribution.

Biology

Most mirids pass the winter in the egg stage. The eggs are laid during the summer in the tender growing stems of the host plant and overwinter. They hatch the following spring when the host plant is sprouting new shoots. The nymphs feed on the new growth, or prey on the arthropods present on the host plant. Each nymph passes through five stages of development, each instar normally taking approximately 5–7 days, and on the fifth molt becomes an adult. The adults feed continually, the males die soon after mating, and the females die soon after they oviposit.

Relatively few species of Miridae hibernate. Adults that hibernate seek shelter late in the fall, usually on the ground or under the loose bark on trees. In the spring they emerge and commence feeding on the available prey or on the tender new shoots. After mating, the females lay eggs and both adults continue to feed throughout the early summer. The eggs incubate for approximately 10–14 days, the nymphs emerge and continue to feed either on the plant or as predators. Each nymph passes through five stages, and becomes an adult. Thus, during the summer the overwintered adults overlap the new generation adults; however, the latter are much more abundant. By midsummer the overwintered adults gradually die out and the new adults continue to feed until hibernation.

Most phytophagous mirids generally are host specific or are limited to a group of closely related plants. However, several are omnivorous and readily disperse to different species of plants. The predaceous mirids are generally found on many different plants preying on the arthropod fauna that these host plants support.

Few species are both predaceous and phytophagous. They have been observed feeding on the foliage or the fruit, and at other times preying on the pest arthropods. Thus, these species are both harmful and beneficial.

Morphology

Fig. 1 shows the typical adult mirid structures and illustrates the structural terms. Adult Miridae collected on fruit crops in Canada are distinguished from the other bugs by the four segmented antennae, the four segmented rostrum, and the absence of ocelli. The hemelytron is typically separated into clavus, corium, embolium, and wing membrane. The abdomen consists of nine segments, but only eight are visible. Each leg consists of coxa, trochanter, femur, tibia, and tarsus. The tarsal claws and the structures between them, the parempodia, and pulvilli provide reliable characters for separating the subfamilies.

The mirid nymphs especially in the early instars are small and delicate, and generally all look alike. Last instar nymphs assume the appearance of the adults except that they do not have fully developed wings, and do not have male or female genital structures. Thus, the identity of nymphs in most situations depends on their association with their adults.

Classification

The classification of the Miridae in this book is the same as that used by me in *The plant bugs of the Prairie Provinces of Canada* (1980c). The subfamilies represented are Mirinae Hahn, Orthotylinae Van Duzee, Phylinae Douglas & Scott, Deraeocorinae Douglas & Scott, and Dicyphinae Reuter.

Key to subfamilies

1. Parempodia large and membranous (Figs. 2,3) 2
   Parempodia slender and hairlike (Figs. 4–8) 3
2. Parempodia divergent toward apices (Fig. 2); pronotal collar distinct (Fig. 9)
   .... **Mirinae** Hahn (p. 13)
   Parempodia parallel or convergent at apices (Fig. 3); pronotal collar depressed, inconspicuous (Fig. 10) ....
   .... **Orthotylinae** Van Duzee (p. 87)
3. Pronotal collar present (Fig. 9) 4
   Pronotal collar absent or depressed (Fig. 10) ....
   .... **Phylinae** Douglas & Scott (p. 120)
4. Pulvilli absent (Fig. 6,7) .... **Deraeocorinae** Douglas & Scott (p. 141)
   Pulvilli present (Fig. 8) .... **Dicyphinae** Reuter (p. 162)

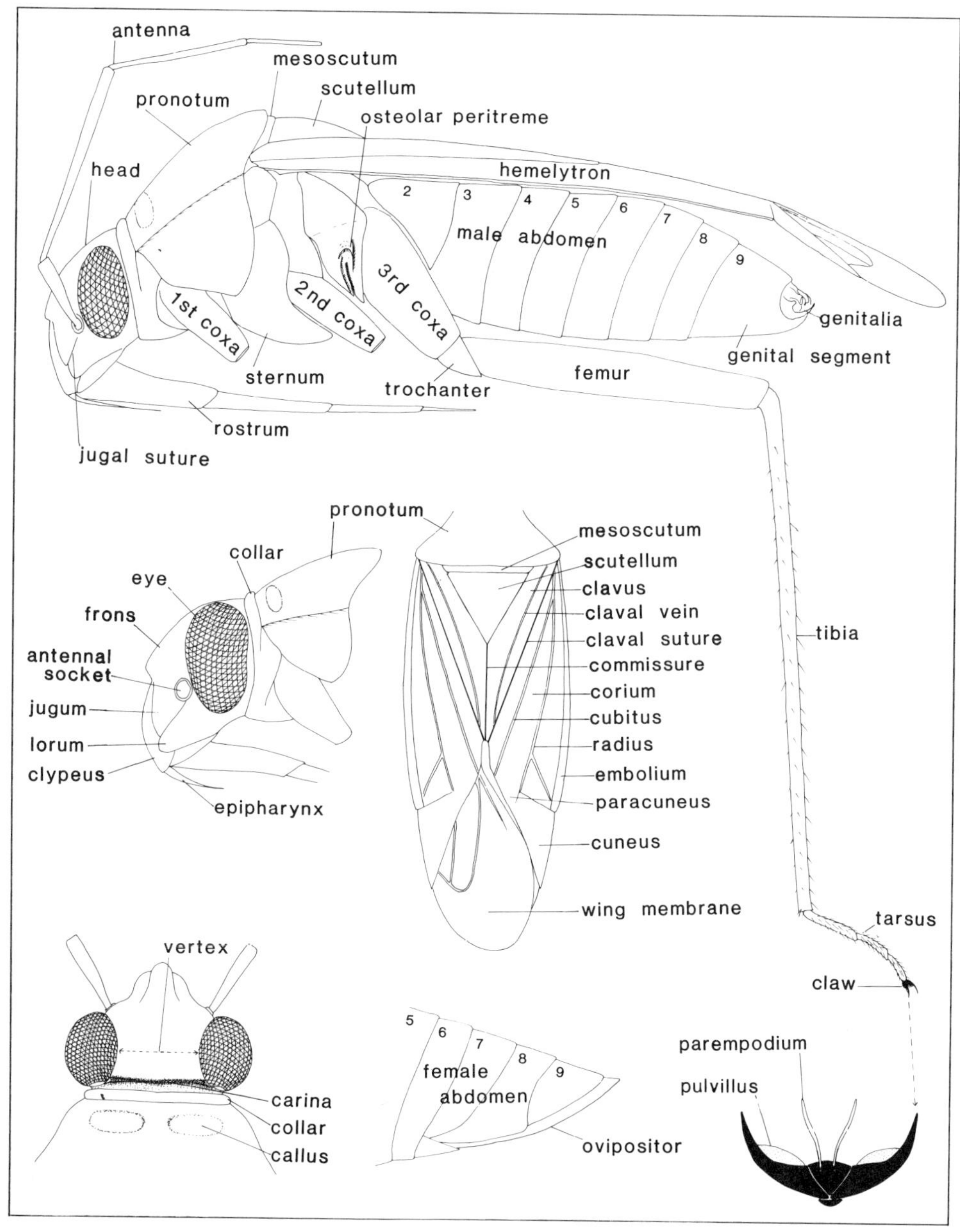

Fig. 1. Adult mirid, showing typical mirid structures and illustrating structural terms.

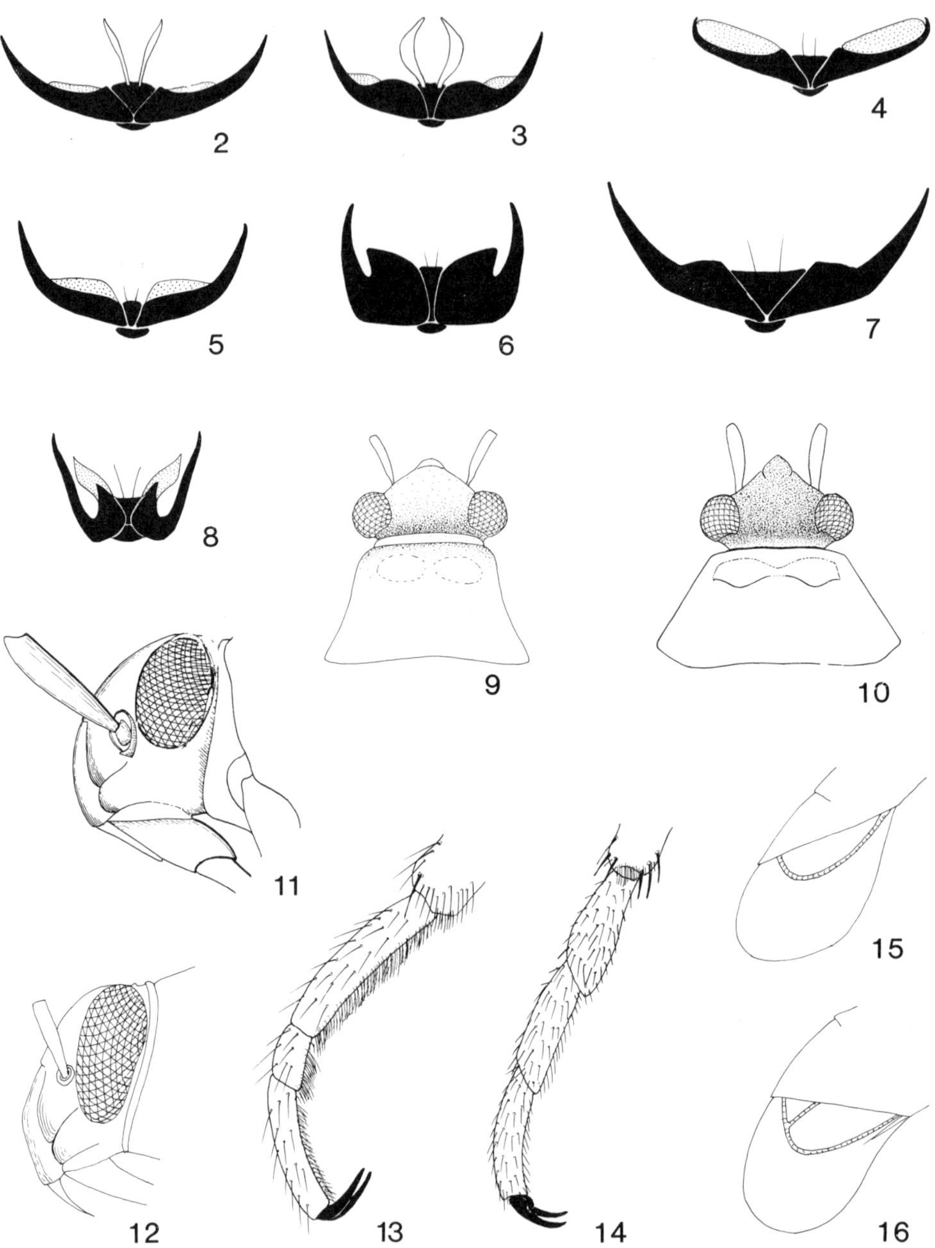

Figs. 2–16. Miridae structures. 2–8, Claws of Miridae; 2, Mirinae; 3, Orthotylinae; 4,5, Phylinae; 6,7, Deraeocorinae; 8, Dicyphinae; 9, Pronotum of Mirinae, Deraeocorinae, and Dicyphinae; 10, Pronotum of Phylinae; 11, Head of *Lygidea*; 12, Head of *Lygus*; 13, Tarsus of *Stenotus*; 14, Tarsus of *Calocoris*; 15, Wing membrane of Hyaliodini; 16, Wing membrane of Deraeocorini.

Subfamily Mirinae Hahn

The following are the subfamily characteristics: 1) large, free parempodia, diverging toward apices; 2) usually prominent pronotal collar; and 3) male genitalia with membranous lobes and flexible ductus seminis.

The subfamily is represented by 2 tribes, 11 genera, and 37 species. Twenty-two species are phytophagous, 15 species are predaceous.

Key to tribes of Mirinae

1. Head, pronotum, and hemelytra velvety . . . . **Resthenini** (p. 13)
   Head, pronotum, and hemelytra shiny, not velvety . . . . **Mirini** (p. 14)

Tribe Resthenini

The tribe is represented by one genus and one species.

Genus *Prepops* Reuter

Elongate, velvety, black and orange species. Head vertical, short. Collar on pronotum prominent. Osteolar peritreme small, indistinct.

One species was collected. Overwinters in the egg stage.

Prepops rubellicollis (Knight)

Fig. 17; Map 1

Platytylellus rubellicollis Knight, 1923*b*:555.
Prepops rubellicollis: Carvalho, 1959:341.

Length 6.6–7.7 mm; width 2.3–2.9 mm. Head black, base orange. Pronotum black, collar and parts of calli orange. Scutellum black. Hemelytra black. Ventral surface and legs black.

Remarks. This species is readily separated from all others by the velvety texture of the dorsal surface, by the large size, and by the orange collar (Fig. 17).

Collected on wild grape in Ontario and Quebec; phytophagous. The nymphs appear about the last part of May and the adults about the end of June. The adults are active throughout July, and gradually die out by the first part of August.

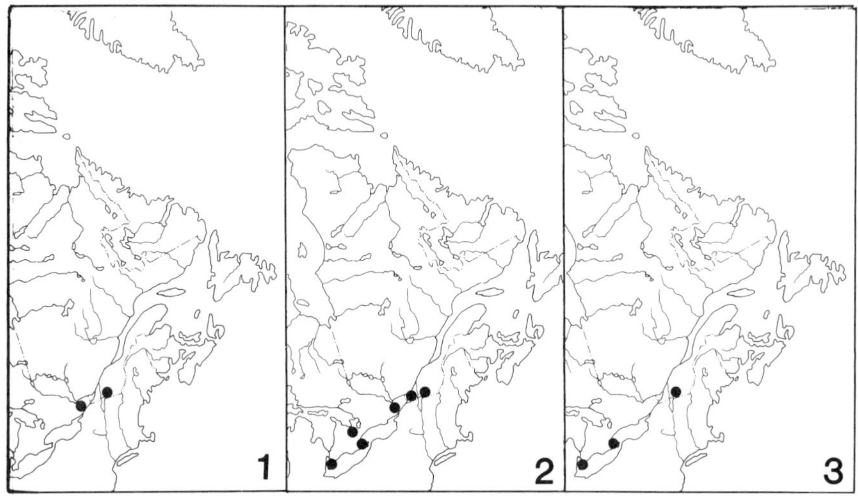

Map 1. Collection localities for *Prepops rubellicollis*.

Map 2. Collection localities for *Neurocolpus nubilus*.

Map 3. Collection localities for *Taedia scrupea*.

Omnivorous, collected on many other herbaceous plants.

Distribution.   Northern half of USA; British Columbia, Prairie Provinces, Ontario, Quebec (Map 1).

Tribe Mirini

The tribe is represented by 10 genera and 36 species. Most of the species are phytophagous, but all 15 species of the genus *Phytocoris* are predaceous.

Key to genera of Mirini

1. First antennal segment with numerous flattened black hairs (Fig. 18) . . . . .
. . . . *Neurocolpus* **Reuter** (p. 15)
   First antennal segment without flattened hairs 2
2. Pronotum with depressed black spots behind each callus (Figs. 19–22) . . . .
. . . . *Taedia* **Distant** (p. 18)
   Pronotum without depressed black spots 3
3. Species black, densely pubescent; second antennal segment clavate (Fig. 23)
. . . . *Capsus* **Fabricius** (p. 24)
   Species not black; second antennal segment linear 4
4. Species with four longitudinal black lines on dorsal surface (Fig. 24) . . . . . .
. . . . *Poecilocapsus* **Reuter** (p. 24)
   Species without four black lines 5

14

Genus *Neurocolpus* Reuter

Robust species. Head oblique; frons elevated and separated from clypeus by deep notch; eyes large, carina between them absent. First antennal segment stout with flattened black hairs. Pronotum and hemelytra impunctate; pubescence silvery, sericeous, intermixed with simple hairs. Legs strongly pilose.

One species was collected. Overwinters in the egg stage.

Neurocolpus nubilus (Say)

Fig. 18; Map 2

Capsus nubilus Say, 1832:22.
Neurocolpus nubilus: Reuter, 1875b:70.

Length 7.0–7.7 mm; width 2.5–2.8 mm. Head light brown; frons often marked with oblique black bars. Rostrum extending to hind coxae. Pronotum yellowish brown with tufts of black, erect hairs intermixed with golden hairs. Hemelytra mottled beige marked with dark brown.

Remarks. The species is distinguished by the flattened black hairs on the first antennal segment (Fig. 18).

Collected on apple in Ontario and Quebec; phytophagous. Also collected on *Rhus typhina*; adults may readily migrate to apple trees if growing nearby. Caesar (1912) and Knight (1922) reported the species as pest of apple in Ontario and New York, respectively.

The nymphs appear in early June and the adults in early July. The adults are common in July, and gradually die out by mid-August.

Distribution. Mexico, widespread in USA; Manitoba, Ontario, Quebec (Map 2).

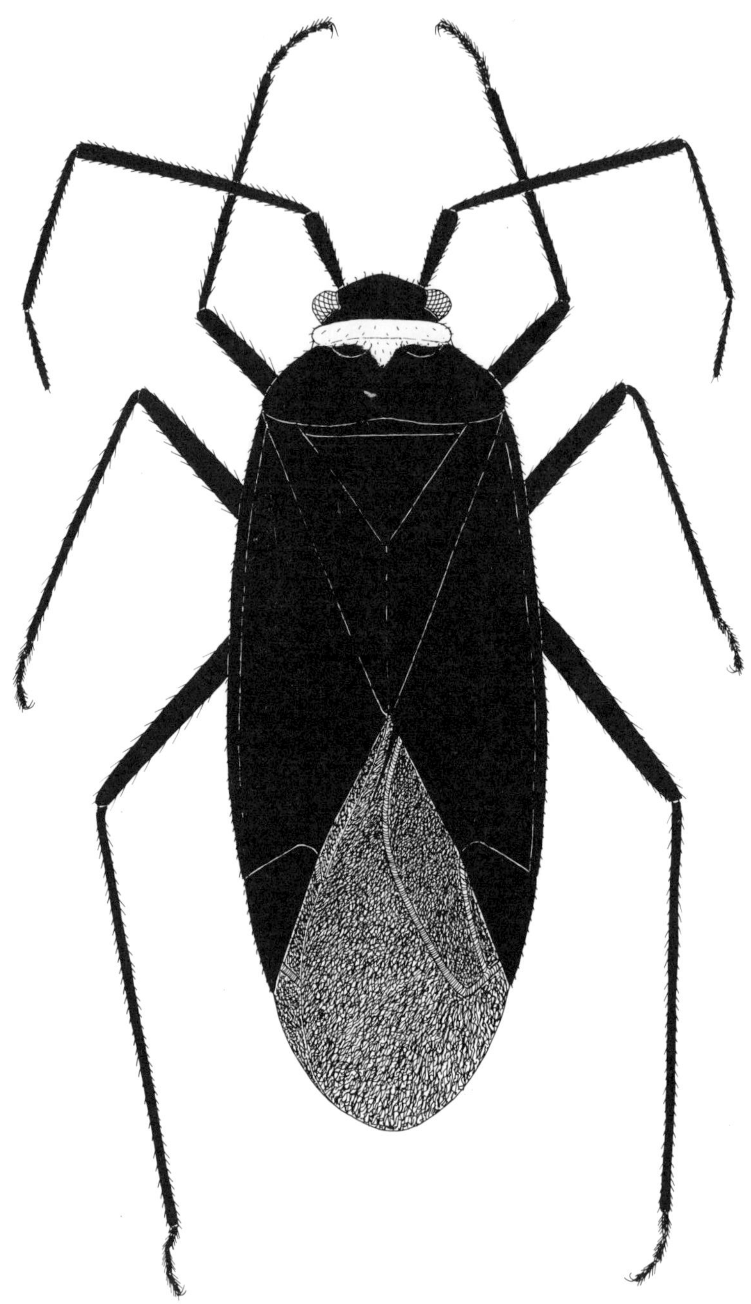

Fig. 17. *Prepops rubellicollis*

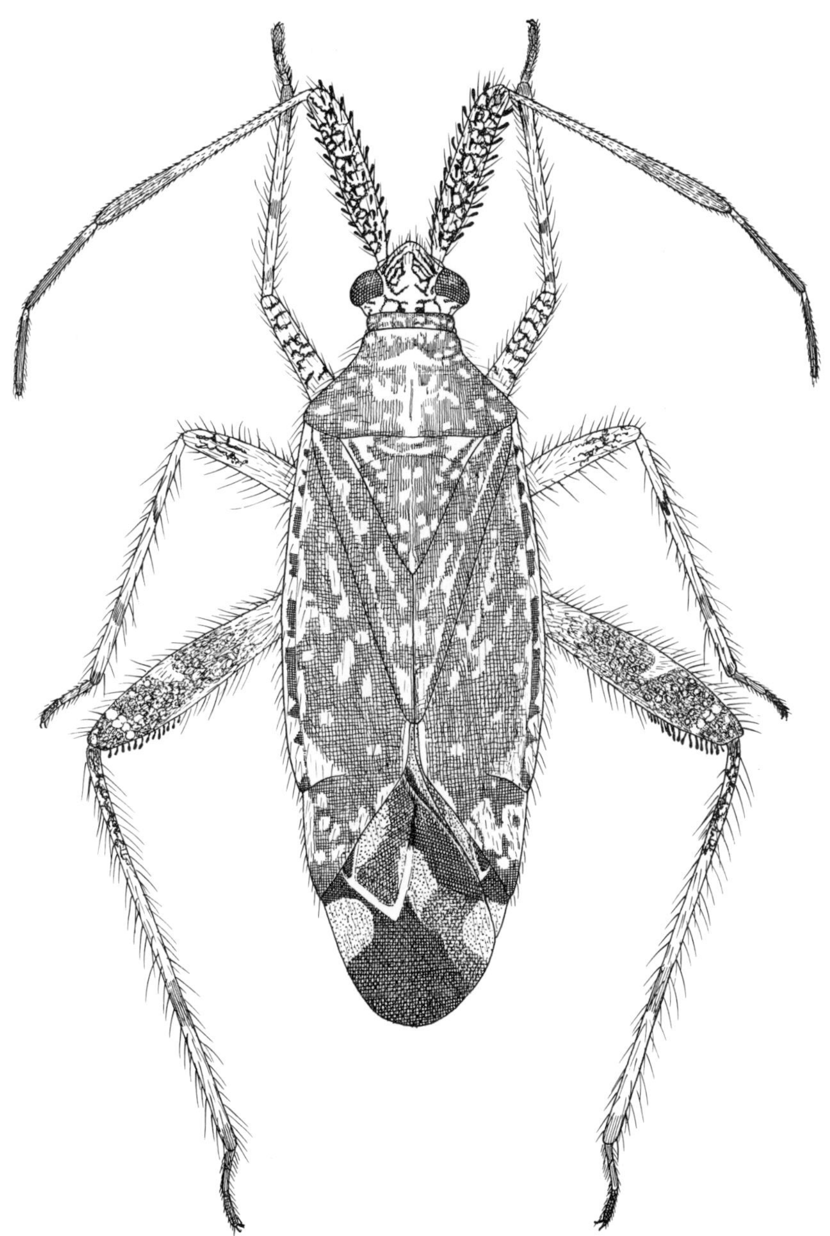

Fig. 18. *Neurocolpus nubilus*

Genus *Taedia* Distant

Elongate, robust species. Head oblique; eyes large, carina between them absent. Pronotum with black velvety spot behind each callus. Hemelytra finely punctate, pubescence sericeous, mixed with simple hairs. Legs long, slender.

Two species were collected. Overwinter in the egg stage. The nymphs appear about mid-May and the adults about mid-June. The adults are common in early July, and gradually die out by the end of July.

Key to species of *Taedia*

1. First antennal segment and tibiae strongly pilose (Figs. 19,20,21) ....
.... *scrupea* (**Say**) (p. 18)
First antennal segment and tibiae not pilose (Fig. 22) ....
.... *pallidula* (**McAtee**) (p. 22)

Taedia scrupea (Say)

Fig. 19; Map 3

Capsus scrupeus Say, 1832:23.
Paracalocoris scrupeus: Reuter, 1909:39.
Taedia scrupeus: Carvalho, 1952:15.

Length 6.3–7.0 mm; width 2.5–2.8 mm. Head yellow marked with brown. First antennal segment black, strongly pilose. Pronotum orange, calli and basal margin often brown. Scutellum orange. Hemelytra black. Tibiae black, strongly pilose.

Remarks. This species is distinguished by the strongly pilose first antennal segments and tibiae. Many varietal names exist; in addition to the typical form *scrupea* (Fig. 19) two other color varieties, *bidens* (Fig. 20) and *varia* (Fig. 21), have been collected.

Collected on wild grape in Ontario and Quebec; phytophagous. McAtee (1916) reported the species on wild apple and wild cherry. Blatchley (1926) reported it on wild grape. Felt (1915) and Knight (1923*b*) reported the species on cultivated grape.

Distribution. Widespread in USA; Ontario, Quebec (Map 3).

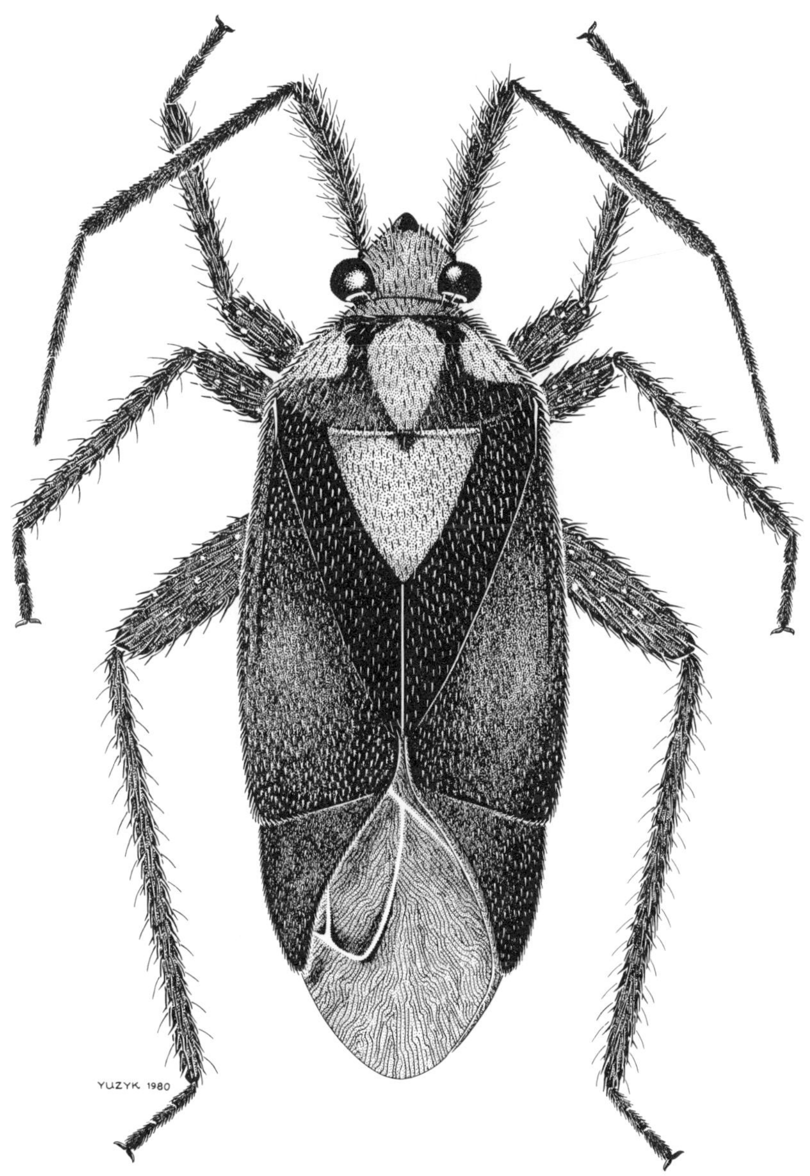

Fig. 19. *Taedia scrupea*

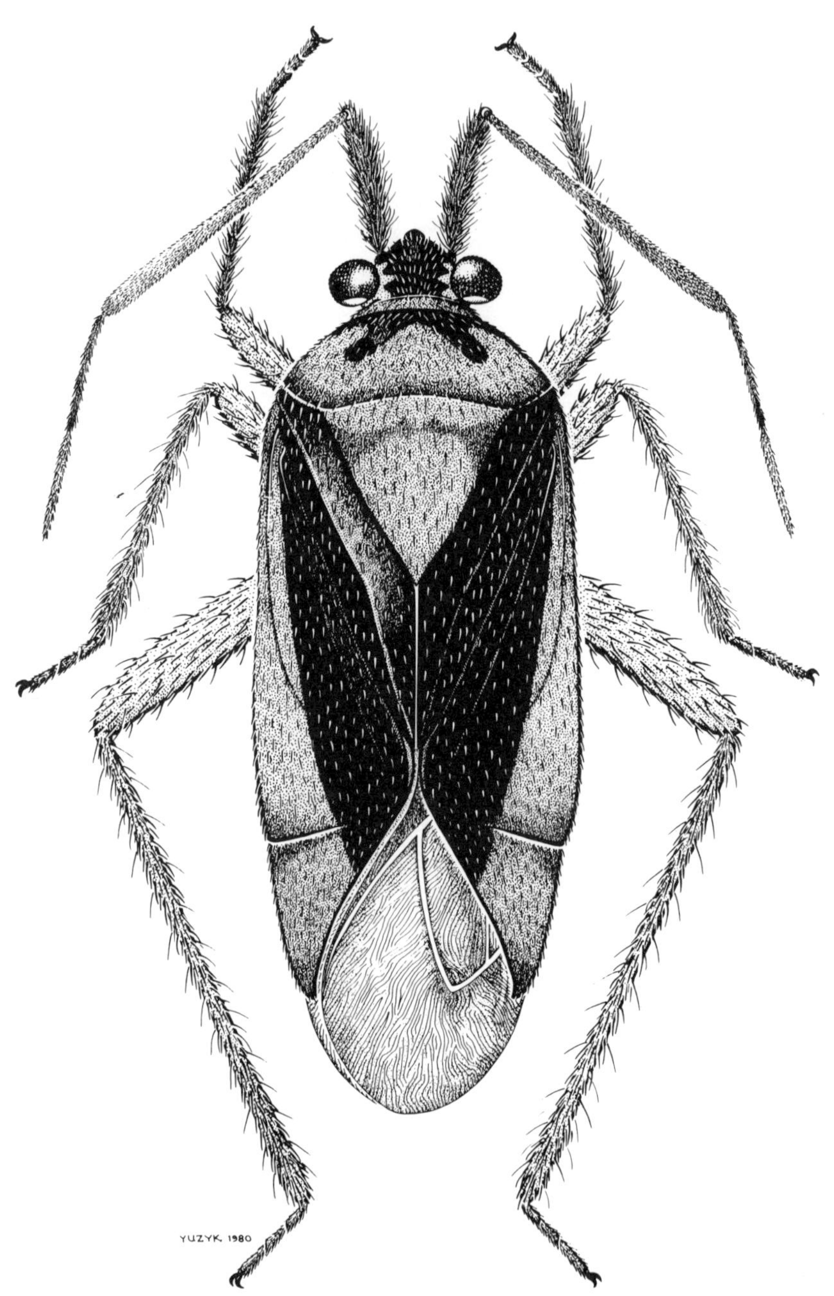

Fig. 20. *Taedia scrupea* var. *bidens*

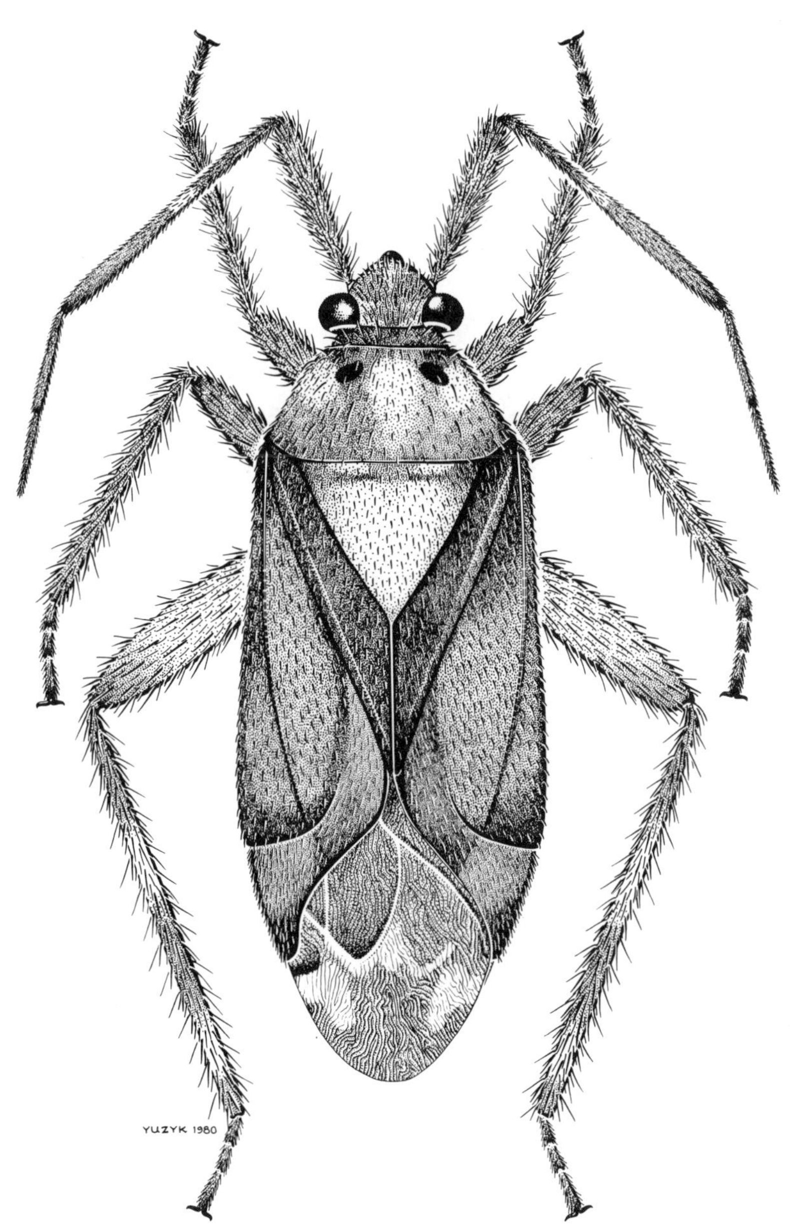

Fig. 21. *Taedia scrupea* var. *varia*

Taedia pallidula (McAtee)

Fig. 22; Map 4

Paracalocoris hawleyi var. *pallidulus* McAtee, 1916:380.
Paracalocoris pallidulus: Knight, 1930*b*:822.
Taedia pallidulus: Carvalho, 1959:262.
Taedia pallidula: Kelton, 1980*c*:57.

Length 6.3–7.0 mm; width 2.3–2.7 mm. Head brown, clypeus darker; frons often marked with oblique black bars. First antennal segment yellow mottled with red. Pronotum brown. Scutellum brown, longitudinal median line yellow. Hemelytra brown with several small yellow dots. Tibiae banded with red.

Remarks. The brown hemelytra with yellow spotting and the absence of pilose hairs on the first antennal segments and tibiae readily distinguish the species (Fig. 22).

Collected on apple in Ontario and Quebec; phytophagous. Caesar (1912) and Knight (1915) referred to *Paracalocoris colon* (Say) as the pest of apple in Ontario and New York, respectively, but later Knight (1922, 1930*b*) confirmed the identity of the species to be *P. pallidula*.

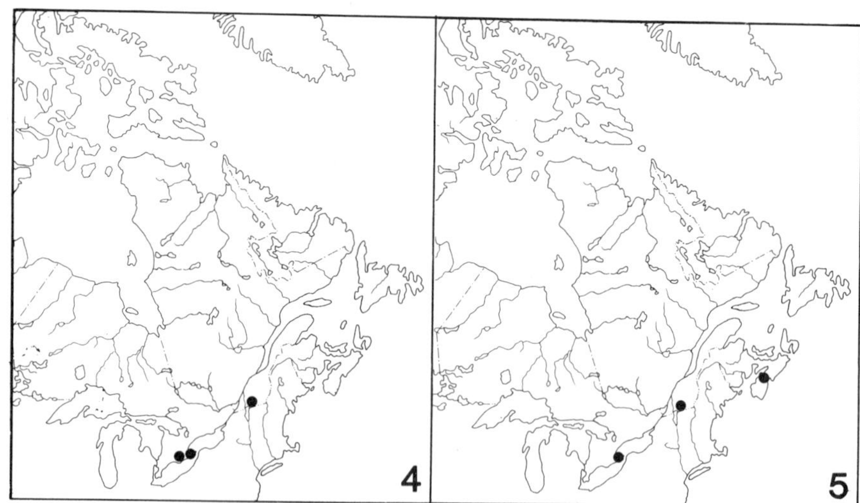

Map 4. Collection localities for *Taedia pallidula*.
Map 5. Collection localities for *Capsus ater*.

Also collected on *Crataegus chrysocarpa*; adults readily migrate from it to apple if the trees are growing nearby.

Distribution.   New York, North Central States, Ohio; Saskatchewan, Manitoba, Ontario, Quebec (Map 4).

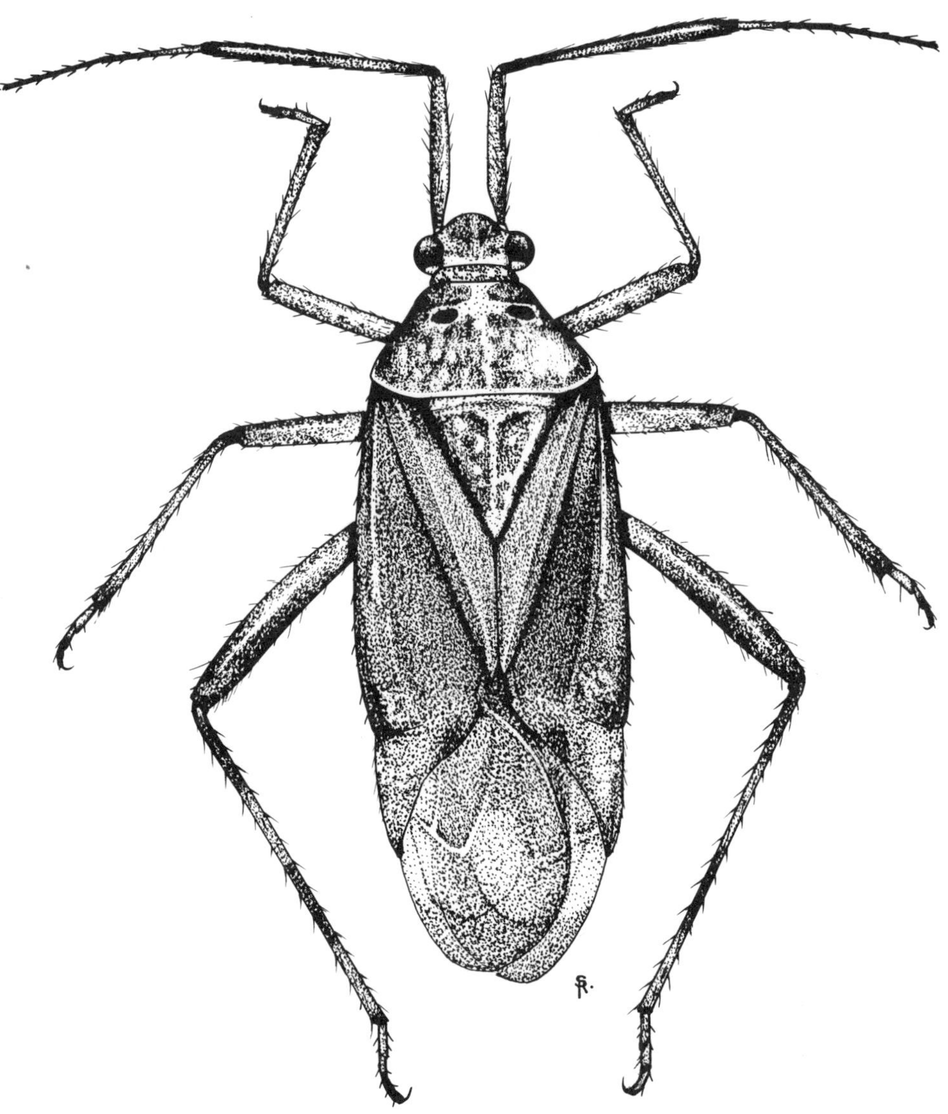

Fig. 22. *Taedia pallidula*

Genus *Capsus* Fabricius

Black, shiny species. Head oblique; carina between eyes absent; second antennal segment clavate. Pronotum and hemelytra punctate; pubescence simple, appressed.

One species was collected. Overwinters in the egg stage.

Capsus ater (Linnaeus)

Fig. 23; Map 5

Cimex ater Linnaeus, 1758:447.
Capsus ater: Fabricius, 1803:241.
Capsus flavipes Provancher, 1872:104.

Length 5.7–6.1 mm; width 2.5–3.0 mm. Head black, area between eyes often pale; second antennal segment clavate. Pubescence on hemelytra simple, silvery, dense.

Remarks.   Provancher (1872) first reported this European species in North America as *Capsus flavipes* and later (1886) as *C. ater*. It is readily distinguished by the black color and the clavate second antennal segment (Fig. 23).

Collected on apple in Nova Scotia, Quebec, Ontario, and British Columbia; phytophagous. The species is known to breed on grasses, but when the grass is cut or during the dry season, the adults readily migrate to the fruit trees and feed on the foliage or fruit.

The nymphs appear in early May and the adults in early June, often earlier. The adults are common throughout June and early July, and die out by the end of July.

Distribution.   Alaska, eastern USA; British Columbia, Ontario, Quebec, Nova Scotia (Map 5).

Genus *Poecilocapsus* Reuter

Glabrous, green with longitudinal black lines. Head vertical, short; carina between eyes absent. Pronotum and hemelytra impunctate, shiny.

One species was collected. Overwinters in the egg stage.

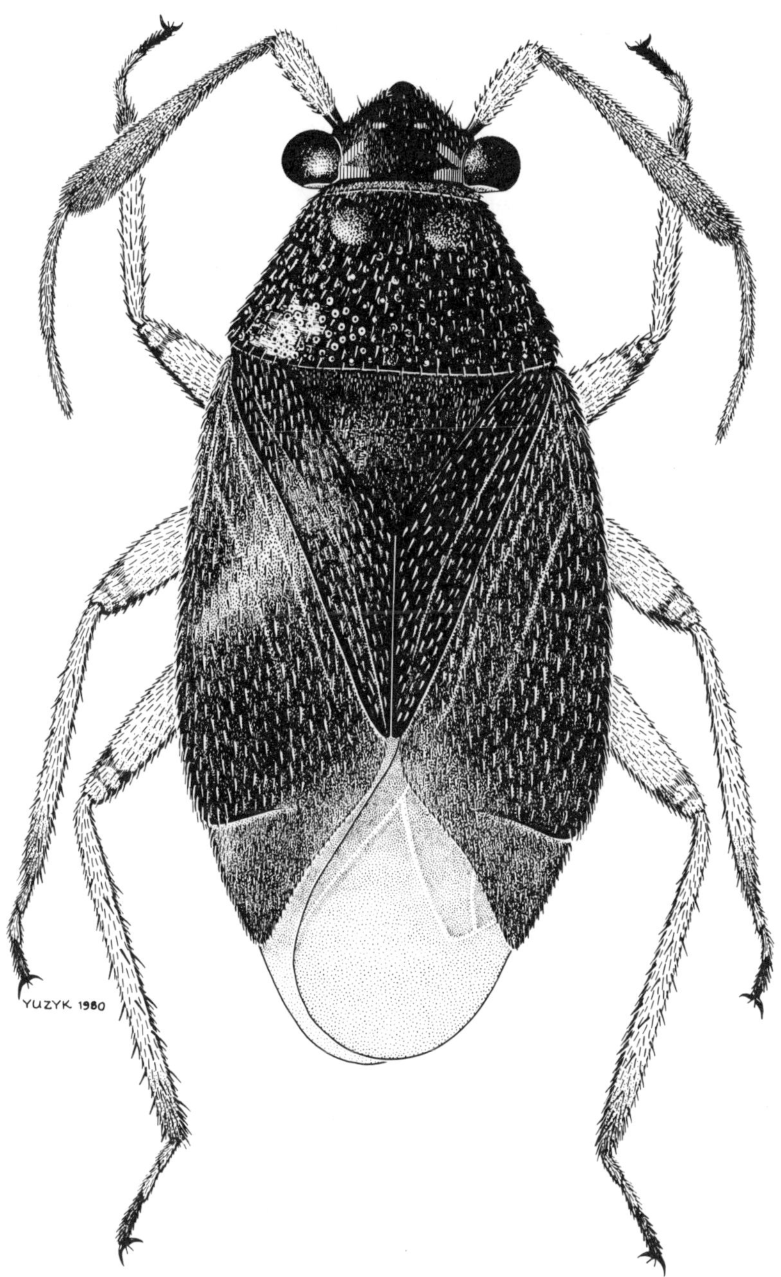

Fig. 23. *Capsus ater*

Poecilocapsus lineatus (Fabricius)

Fig. 24; Map 6

Lygaeus lineatus Fabricius, 1798:451.
Capsus quadrivittatus Say, 1832:20.
Phytocoris bellus Emmons, 1854:30.
Poecilocapsus lineatus: Reuter, 1875*b*:74.

Length 7.0–7.5 mm; width 2.8–3.5 mm. Head brown; clypeus and antennae black. Pronotum and hemelytra yellowish green with four black longitudinal lines. Legs green.

Remarks. This species is distinguished by the yellowish green color with four black lines on the dorsum (Fig. 4).

Collected on raspberry in New Brunswick and Nova Scotia; on raspberry and wild grape in Ontario; phytophagous. Lochhead (1903) reported the species as a pest of raspberry and Gibson (1905) as a pest of currant in Ontario.

Also collected on *Mentha arvensis* and many other plants, including potato.

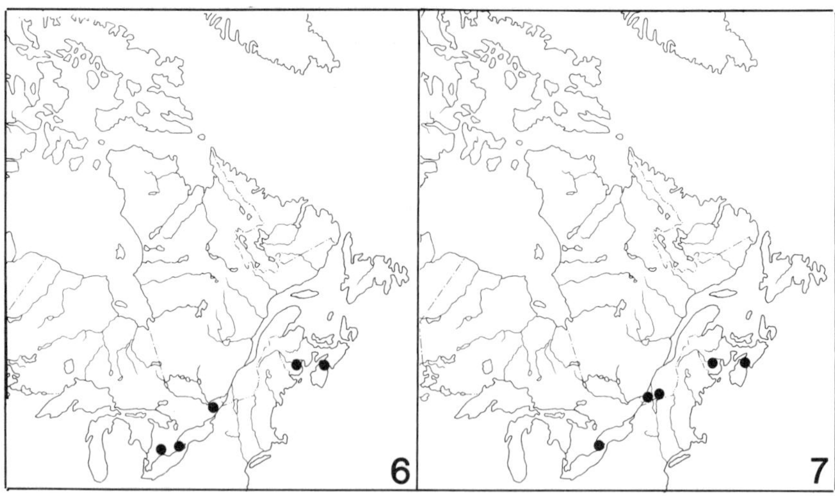

Map 6. Collection localities for *Poecilocapsus lineatus*.
Map 7. Collection localities for *Lygidea mendax*.

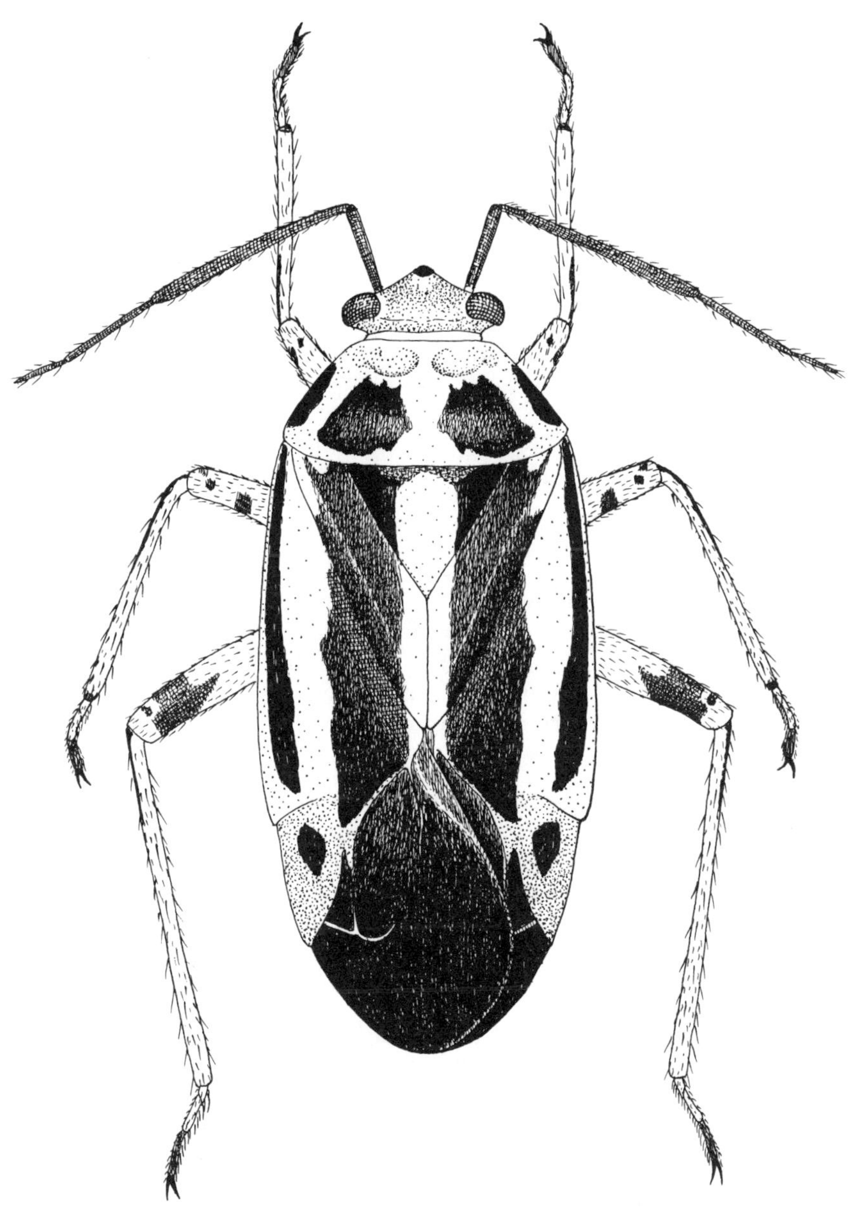

Fig. 24. *Poecilocapsus lineatus*

Nymphs appear about mid-May and the adults about mid-June. The adults are common from mid-June to mid-July, and gradually die out by mid-August.

Distribution. Widespread in USA; Nova Scotia, New Brunswick, Quebec, Ontario, Manitoba, Saskatchewan (May 6).

Genus *Lygidea* Reuter

Elongate, reddish brown species. Head nearly vertical; frons smooth; eyes nearly spherical positioned above antennal sockets; carina between eyes distinct. Pronotum coarsely punctate, calli smooth. Hemelytra coarsely punctate, pubescence simple, dense.

One species was collected. Overwinters in the egg stage.

Lygidea mendax Reuter

Figs. 11, 25; Map 7

Lygidea mendax Reuter, 1909:47.

Length 6.2–6.5 mm; width 2.1–2.3 mm. Head red or orange; clypeus and antennae black. Pronotum coarsely punctate, orange red; narrow basal submargin black. Scutellum red marked with black. Hemelytra mostly brown, coastal margins orange red; pubescence golden. Ventral surface red, hind tibia black.

Remarks. This species is distinguished by the coarsely punctate pronotum and hemelytra, by the red or orange head, pronotum, and coastal margins on the hemelytra (Fig. 25).

Collected on apple in Nova Scotia, New Brunswick, Quebec, and Ontario; phytophagous. This species was reported by Caesar (1912) as a pest of apple in Ontario, by Brittain (1915a) as a pest of apple in Nova Scotia; by MacNay (1962) as a pest of apple in New Brunswick; and by Rivard and Paradis (1978) as a pest of apple in Quebec. Knight (1915, 1941b) reported it as a serious pest of apple in New York, Pennsylvania, and Michigan.

The nymphs appear about mid-May and the adults about mid-June. The adults are common from mid-June to mid-July, and gradually die out by the end of July.

Also breeds on *Crataegus chrysocarpa*, and the adults readily migrate from it to apple trees if grown nearby, and damage the fruit.

Distribution. Northeastern and north central USA; Nova Scotia, New Brunswick, Quebec, Ontario (Map 7).

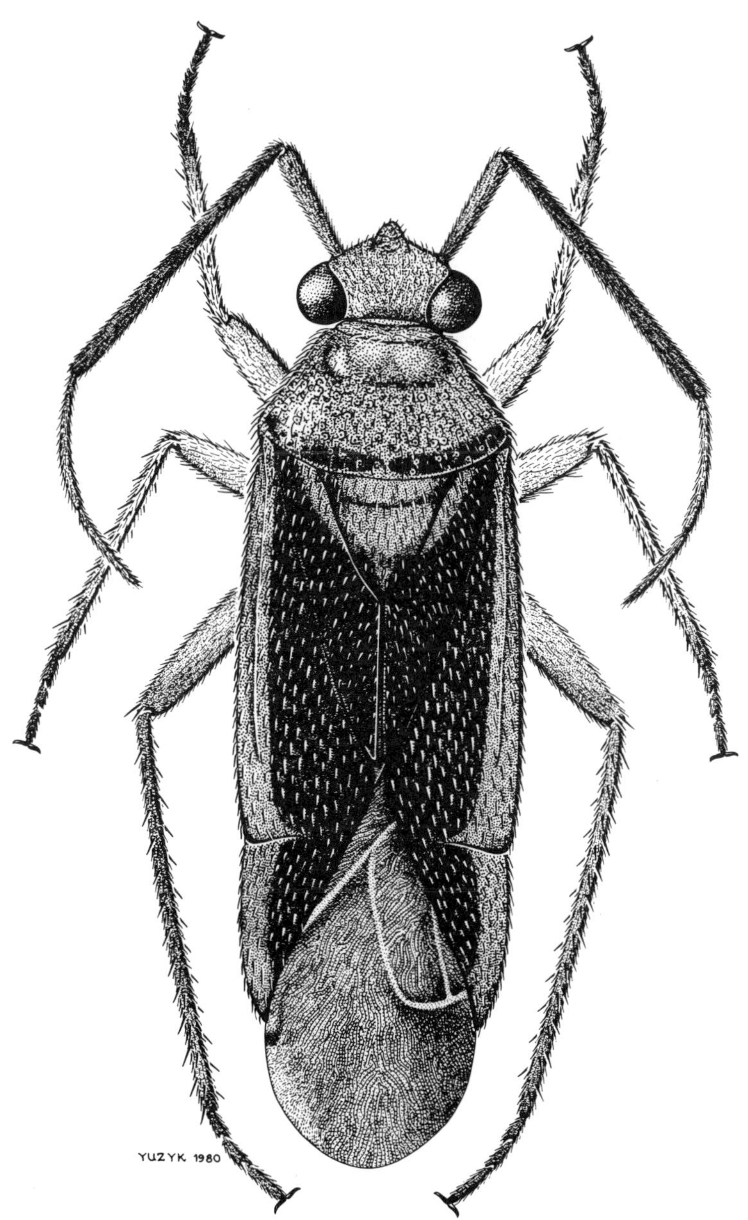

Fig. 25. *Lygidea mendax*

Genus *Lygus* Hahn

Elongate-oblong, reddish brown species. Head oblique; eyes large, carina between them prominent. Pronotum and hemelytra coarsely punctate.

Six species were collected. The adults hibernate. The adults emerge early in the spring and commence feeding on the available plants. After mating, the females oviposit throughout May, June, and July. The first nymphs appear about the end of May, the majority in June and July. The new generation adults appear about the end of June. The population is at the maximum at the end of June when the overwintered adults, nymphs, and new generation adults are present together, and at this time cause maximum damage to fruit. By August most of the overwintered adults die out; the new generation adults continue to feed until hibernation.

For other species in North America see Kelton (1975).

Key to species of *Lygus*

1. Frons striate or grooved (Fig. 26) .... *nubilus* **Van Duzee** (p. 30)
   Frons smooth, not striate or grooved 2
2. Frons with submedian oblique bars (Figs. 27,28) 3
   Frons without submedian oblique bars (Figs. 29–31) 4
3. Pubescence on hemelytra uniformly yellow (Fig. 27) ....
   .... *lineolaris* **(Palisot de Beauvois)** (p. 31)
   Pubescence on hemelytra with patches of silvery hairs (Fig. 28) ....
   .... *plagiatus* **Uhler** (p. 34)
4. Pubescence on hemelytra long and dense (Fig. 29) ....
   .... *hesperus* **Knight** (p. 36)
   Pubescence on hemelytra short and sparse 5
5. Anterior angles of pronotum prominent (Fig. 30) .. *varius* **Knight** (p. 39)
   Anterior angles of pronotum rounded (Fig. 31) .... *shulli* **Knight** (p. 41)

Lygus nubilus Van Duzee

Fig. 26; Map 8

Lygus distinguendus var. *nubilus* Van Duzee, 1914:20.
Lygus nubilus Van Duzee, 1917:350.
Lygus ultranubilus Knight, 1917b:583.
Lygus epelys Hussey, 1954:196.

Length 4.3–5.5 mm; width 1.8–2.3 mm. Head yellowish brown; frons striate. Rostrum 1.5–2.0 mm long. Pronotum yellowish brown; calli pubescent. Mesoscutum yellowish or light reddish. Hemelytra yellowish mottled with dark brown; pubescence long, dense.

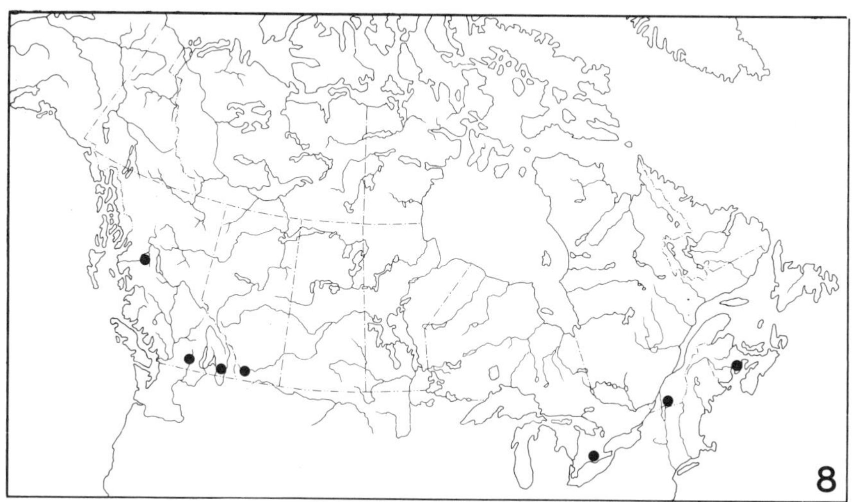

Map 8. Collection localities for *Lygus nubilus*.

Remarks.   This species is distinguished by the small size, short rostrum, pubescent calli, and striate frons (Fig. 26).

Collected on elderberry in British Columbia, Alberta, Ontario, and Quebec; phytophagous.

Distribution.   Western USA, Michigan, Connecticut; British Columbia, Alberta, Ontario, Quebec, New Brunswick (Map 8).

Lygus lineolaris (Palisot de Beauvois)

Fig. 27; Map 9

Capsus lineolaris Palisot de Beauvois, 1818:187.
Lygus oblineatus Say, 1832:21.
Capsus flavonotatus Provancher, 1872:103.
Lygus lineolaris: Uhler, 1872:413.

Length 4.9–5.9 mm; width 2.3–3.0 mm. Head yellowish brown; frons with red or black submedian oblique bars. Mesoscutum black, lateral areas pale or reddish. Hemelytra yellowish or reddish brown; pubescence yellow, long, dense.

Remarks.   This species is distinguished by the submedian oblique bars on the frons, by the pale or reddish lateral areas on the mesoscutum, and by the yellow, long, and dense pubescence on the hemelytra (Fig. 27).

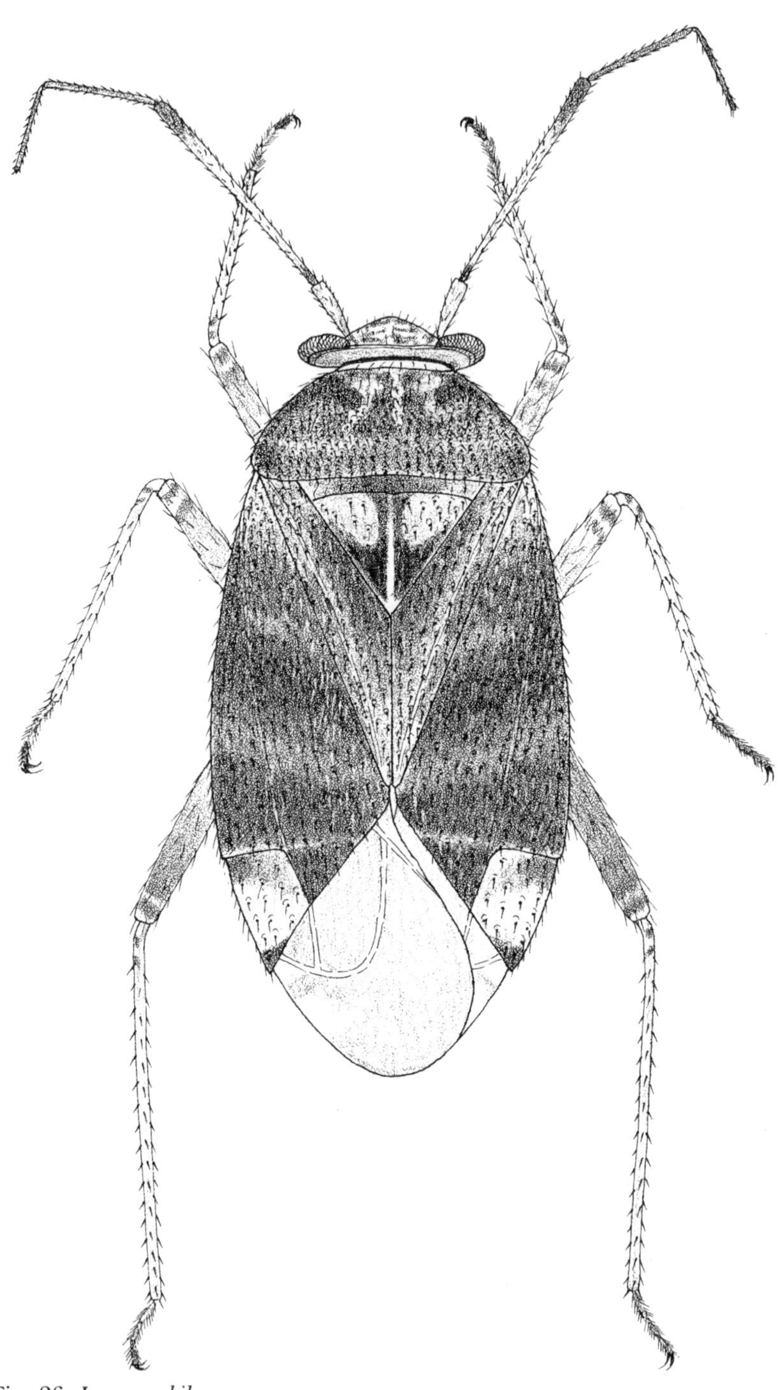

Fig. 26. *Lygus nubilus*

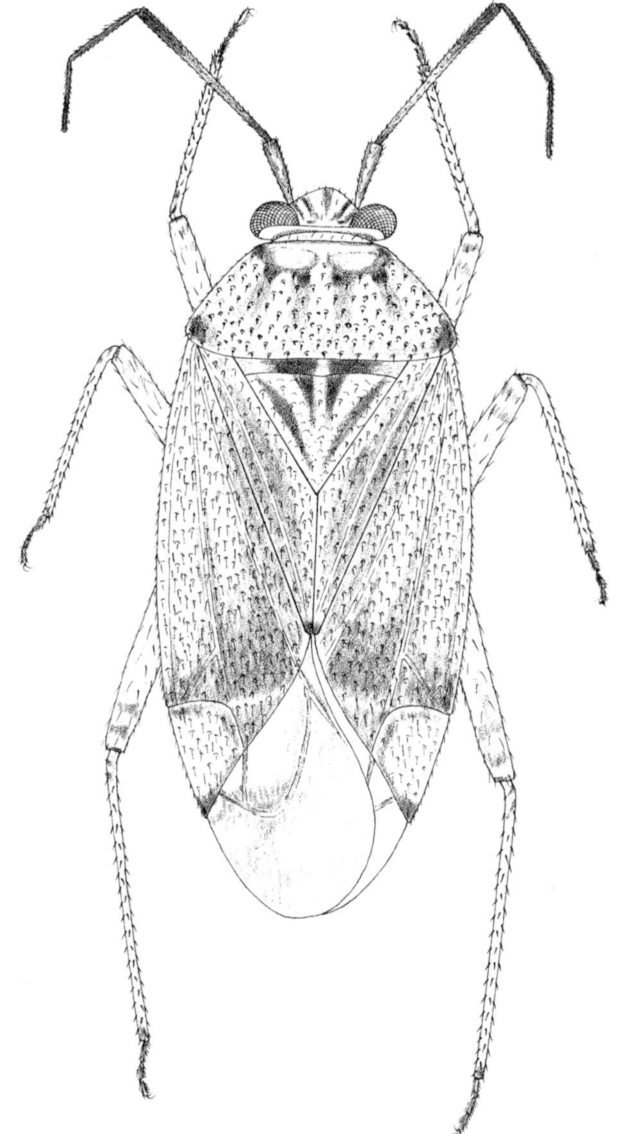

Fig. 27. *Lygus lineolaris*

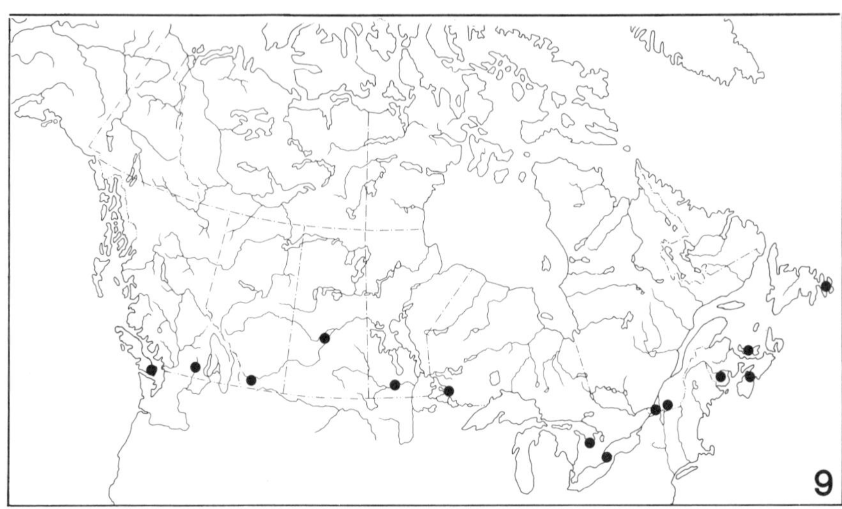

Map 9. Collection localities for *Lygus lineolaris.*

This is the familiar tarnished plant bug and it is the most common and most omnivorous pest of fruit crops encountered in Canada (Table 1). The literature on the pest is voluminous; for key references see Kelton (1975).

Distribution.   Widespread in USA; all provinces of Canada (Map 9).

Lygus plagiatus Uhler

Fig. 28; Map 10

Lygus plagiatus Uhler, 1895:35.

Length 5.0–5.9 mm; width 2.5–2.9 mm. Head light brown; clypeus, lorum, and jugum marked with black; frons with submedian oblique black bars. Mesoscutum black, lateral areas red. Hemelytra dull green mottled with black; pubescence long and dense, yellow and white.

Remarks.   The markings on the frons are similar to those of *lineolaris*, but the mottled appearance of the hemelytra readily distinguish this species (Fig. 28).

Collected on peach and pear in Ontario; phytophagous. The species is not as omnivorous as *lineolaris* and is usually found on *Ambrosia trifida*.

Distribution.   North central and northeastern USA; Prairie Provinces, Ontario, Quebec (Map 10).

34

Table 1. Tarnished plant bug on fruit crops

	B.C.	Alta.	Sask.	Man.	Ont.	Que.	N.B.	N.S.	P.E.I.	Nfld.
apple	x				x	x	x	x	x	
pear	x				x	x		x		
peach	x				x					
plum	x				x	x			x	
apricot	x				x					
sweet cherry	x				x					
sour cherry	x				x					
black cherry					x	x	x	x		
pin cherry		x	x	x	x	x	x	x	x	x
chokecherry	x	x	x	x	x	x	x	x	x	x
raspberry	x	x	x	x	x	x	x	x	x	x
blackberry	x									
thimbleberry	x				x	x	x	x		
loganberry	x									
currant	x	x	x	x	x	x	x	x	x	x
gooseberry	x	x	x	x	x	x	x	x	x	x
serviceberry	x	x	x	x	x	x	x	x	x	x
cranberry	x		x	x	x	x	x	x		
viburnum	x	x	x	x	x	x	x	x	x	x
strawberry	x	x	x	x	x	x	x	x	x	x
blueberry			x	x	x	x	x	x	x	x
grape	x				x	x				
elderberry	x	x			x	x	x			
mulberry					x					

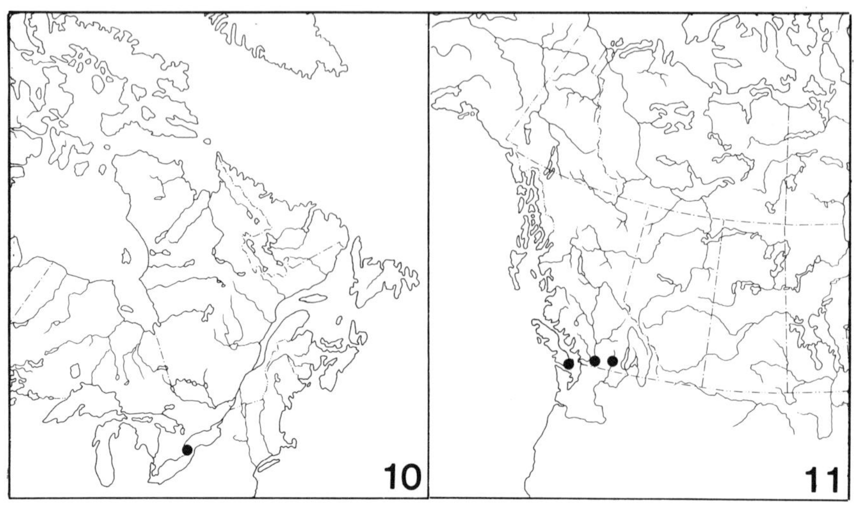

Map 10. Collection locality for *Lygus plagiatus.*

Map 11. Collection localities for *Lygus hesperus.*

Lygus hesperus Knight

Fig. 29; Map 11

Lygus elisus var. *hesperus* Knight, 1917*b*:575.
Lygus hesperus: Shull, 1933:1.

Length 5.3–6.5 mm; width 2.3–2.8 mm. Head yellowish green. Mesoscutum black. Hemelytra yellowish green, apical half of corium often red; pubescence long, dense.

Remarks. This species is distinguished by the clear frons, and by the long and dense pubescence on the hemelytra (Fig. 29).

Collected on apple, pear, sweet cherry, plum, peach, raspberry, blackberry, loganberry, and currant in British Columbia; phytophagous. Twinn (1939) reported it as a pest of pear and peach in British Columbia. It also feeds on a great variety of other plants including alfalfa and vegetable crops.

The damage to fruit crops is similar to that of *lineolaris*. It is perhaps the most common species of *Lygus* in areas where agriculture is carried on in British Columbia.

Distribution. Western USA; British Columbia (Map 11).

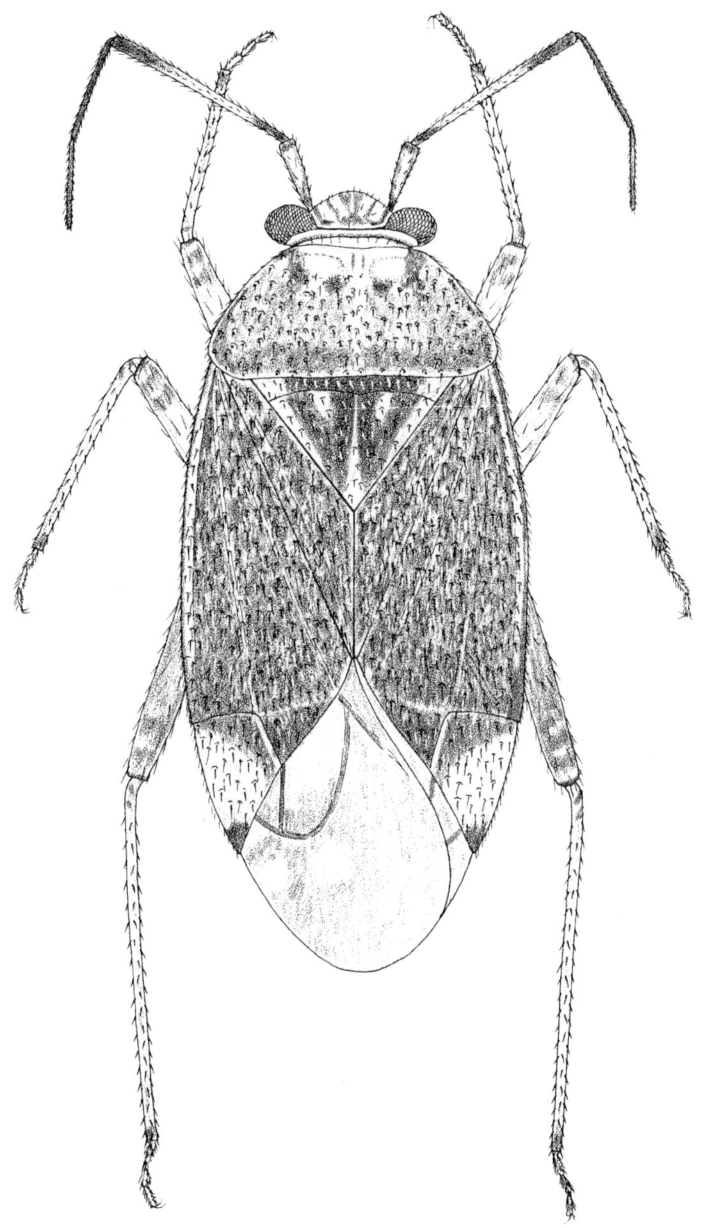

Fig. 28. *Lygus plagiatus*

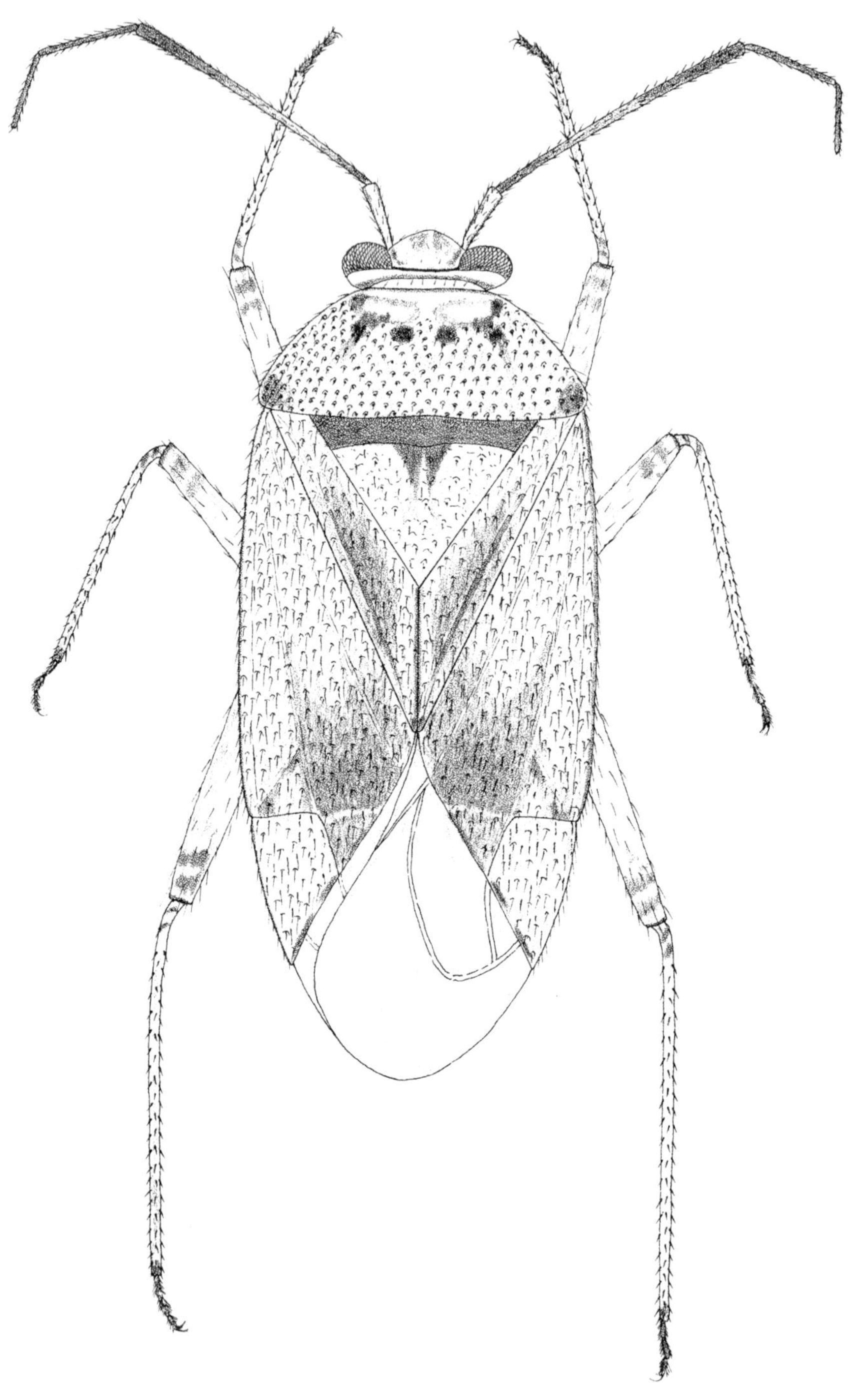

Fig. 29. *Lygus hesperus*

Lygus varius Knight

Fig. 30; Map 12

Lygus varius Knight, 1944:473.

Length 5.6–5.8 mm; width 2.7–3.0 mm. Head light yellowish brown; frons with black or reddish brown inverted "V". Mesoscutum black. Hemelytra greenish brown to dark brown; pubescence short, sparse.

Remarks. This species resembles *lineolaris* in size and color but is easily separated from it by the inverted "V" on the frons, by the black mesoscutum, and by the short and sparse pubescence on the hemelytra (Fig. 30).

Cram (*in litt.*) observed the adults on strawberry in British Columbia where they caused severe fruit deformity.

Also collected on many other plants.

Distribution. Western USA; Newfoundland, Quebec, Ontario, Saskatchewan, Alberta, British Columbia (Map 12).

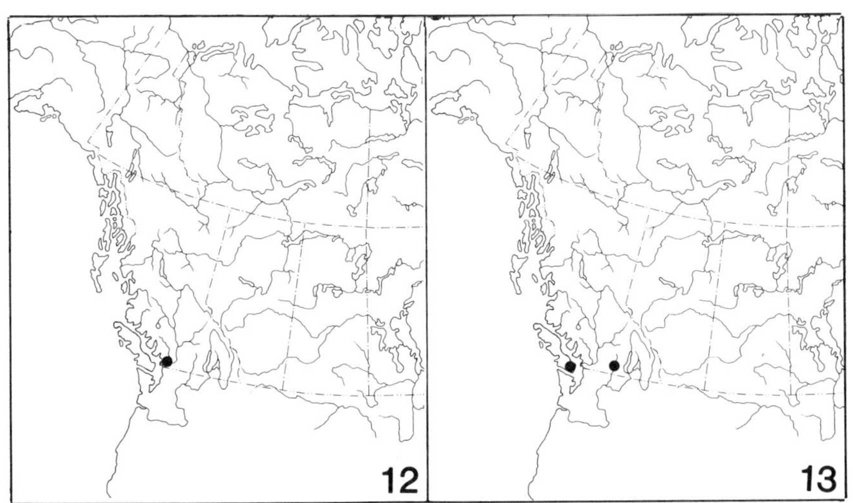

Map 12. Collection locality for *Lygus varius*.

Map 13. Collection localities for *Lygus shulli*.

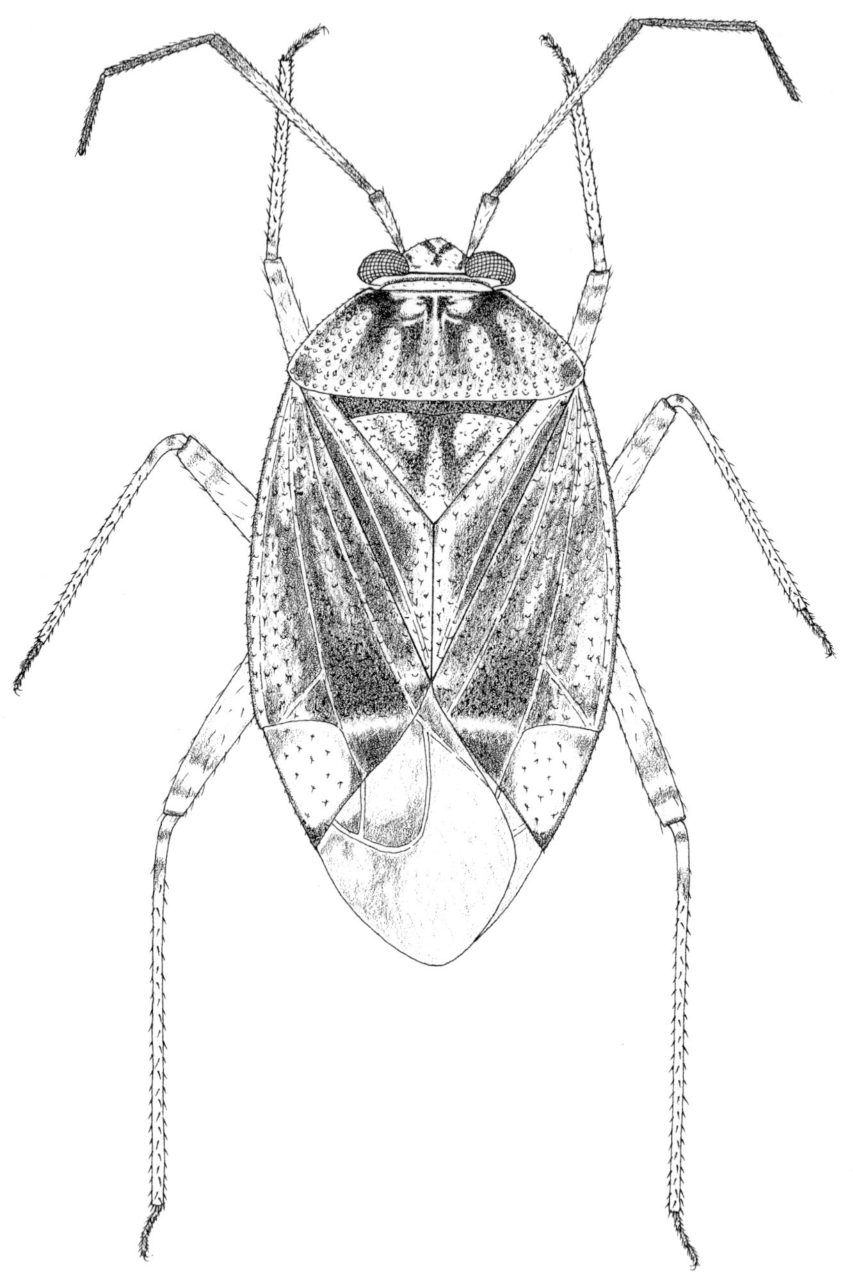

Fig. 30. *Lygus varius*

Lygus shulli Knight

Fig. 31; Map 13

Lygus shulli Knight, 1941*a*: 272.

Length 5.4–6.0 mm; width 2.6–3.0 mm. Head yellowish brown. Pronotum yellowish or greenish brown. Mesoscutum black. Hemelytra greenish yellow; middle of clavus and apical area of corium dark brown; pubescence short, sparse.

Remarks. This species is easily confused with *hesperus* (Fig. 29) as both look alike and have similar markings, but *shulli* has shorter and sparser pubescence (Fig. 31).

Collected on peach, sweet cherry, blackberry, loganberry, and thimbleberry in British Columbia; phytophagous. Twinn (1938) probably referred to *shulli* as a pest of pear, and Buckell (1939) as a pest of peach. Knight (1941*a*) reported the species as a pest of peach in Washington, USA.

Also collected on many other plants.

Distribution. Western USA; Prairie Provinces, British Columbia (Map 13).

Genus *Lygocoris* Reuter

Elongate-oblong, green, or green and black species. Head oblique; eyes large, carina between them prominent. Pronotum and hemelytra finely punctate; pubescence simple, long, dense.

Eight species were collected. Overwinter in the egg stage. The nymphs appear in early May, sometimes earlier, and adults about the first part of June. The adults are short-lived, and after mating, the females oviposit in the tender new growth; by the end of July most of them die out. Thus, most of the damage to fruit crops is done by nymphs in May and by adults in June and early July.

For other species in North America see Kelton (1971*b*).

Key to species of *Lygocoris*

1. Head, pleura, abdomen, and hind femora strongly marked with red; pronotum with dark ray behind each callus (Fig. 32); male claspers (Fig. 40) . . . . **communis (Knight)** (p. 42)
   Head, pleura, abdomen, and hind femora green or marked with black .. 2
2. Pronotum, scutellum, and hemelytra mostly black (Fig. 33); male claspers (Fig. 41) . . . . **caryae (Knight)** (p. 43)

Pronotum, scutellum, and hemelytra mostly green, or with black areas (Figs.
34–37) 3

3. Ventral surface mostly green 4
   Ventral surface mostly brown or black 5

4. Rostrum 1.7 mm or shorter; male claspers (Fig. 42) ....
   .... ***inconspicuus*** **(Knight)** (p. 48)
   Rostrum 1.8 mm or longer; male claspers (Fig. 43) ....
   .... ***belfragii*** **(Reuter)** (p. 50)

5. Second antennal segment black (Fig. 36); male claspers (Fig. 44) ....
   .... ***knighti*** **Kelton** (p. 52)
   Second antennal segment pale green, apex black (Figs. 37–39) 6

6. Rostrum shorter than 1.7 mm; male claspers (Fig. 45) ....
   .... ***viburni*** **(Knight)** (p. 54)
   Rostrum 1.7 mm or longer 7

7. Pronotum with pale green calli (Fig. 38); male claspers (Fig. 46) ....
   .... ***omnivagus*** **(Knight)** (p. 54)
   Pronotum with dark calli (Fig. 39); male claspers (Fig. 47) ....
   .... ***quercalbae*** **(Knight)** (p. 56)

Lygocoris communis (Knight)

Figs. 32, 40; Map 14

Lygus communis Knight, 1916:346.
Neolygus communis: Knight, 1941*b*:159.
Lygocoris (Neolygus) communis: Carvalho, 1959:141.

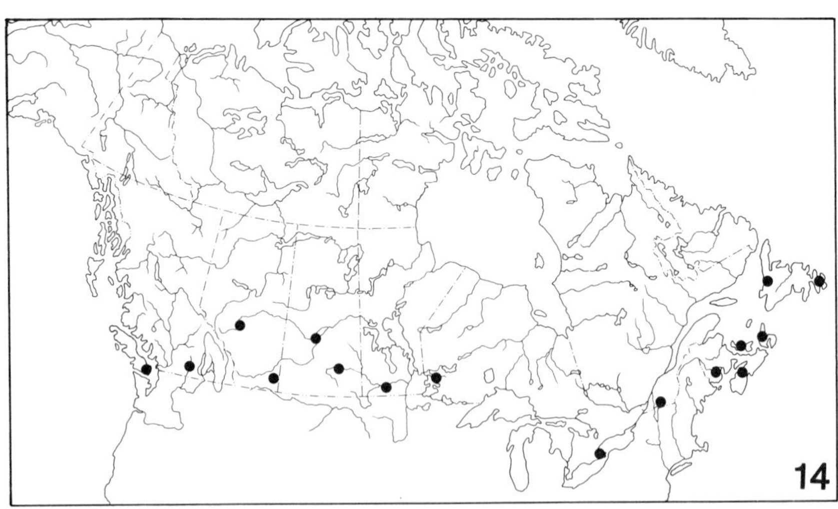

Map 14. Collection localities for *Lygocoris communis*.

Length 5.1–5.9 mm; width 2.2–2.6 mm. Head yellowish marked with transverse reddish bars. Pronotum yellowish green, ray behind each callus reddish or black. Scutellum yellowish, median line usually reddish. Hemelytra mostly reddish brown. Ventral surface greenish, pleuron and side of abdomen reddish. Femora marked with red.

Remarks.   This species is commonly known as the pear plant bug. It is distinguished by the reddish bars on the frons, by the reddish or black rays behind the calli (Fig. 32), by the reddish pleuron, abdomen, and hind femora, and by the claspers (Fig. 40).

This species is an important pest of fruit crops like the tarnished plant bug. It was collected on all fruit crops except strawberry and blueberry (Table 2). Brittain (1915*b*) reported the species (cited as *Lygus invitus*) as a pest of apple in Nova Scotia, which Knight (1916) described as *communis*.

Distribution.   Transcontinental in USA; British Columbia, Prairie Provinces, Ontario, Quebec, Atlantic Provinces (Map 14).

Lygocoris caryae (Knight)

Figs. 33, 41; Map 15

Lygus (Neolygus) caryae Knight, 1917*b*:161.
Neolygus caryae Knight, 1941*b*:161.
Lygocoris (Neolygus) caryae: Carvalho, 1959:141.

Length 4.9–5.8 mm; width 1.9–2.4 mm. Head dark brown; second antennal segment black. Pronotum, scutellum, and hemelytra mostly black.

Remarks.   This species is commonly known as the hickory plant bug. It is distinguished by the black color (Fig. 33).

Collected on peach and apricot in Ontario; phytophagous. Caesar (1920) and Knight (1941*b*) reported the species "catfacing" peach in Ontario, and in New York and Ohio, respectively.

Breeds on *Carya ovata*; adults readily migrate to orchard trees and feed on the fruit, especially if the fruit trees are nearby.

Distribution.   Eastern USA; Quebec, Ontario (Map 15).

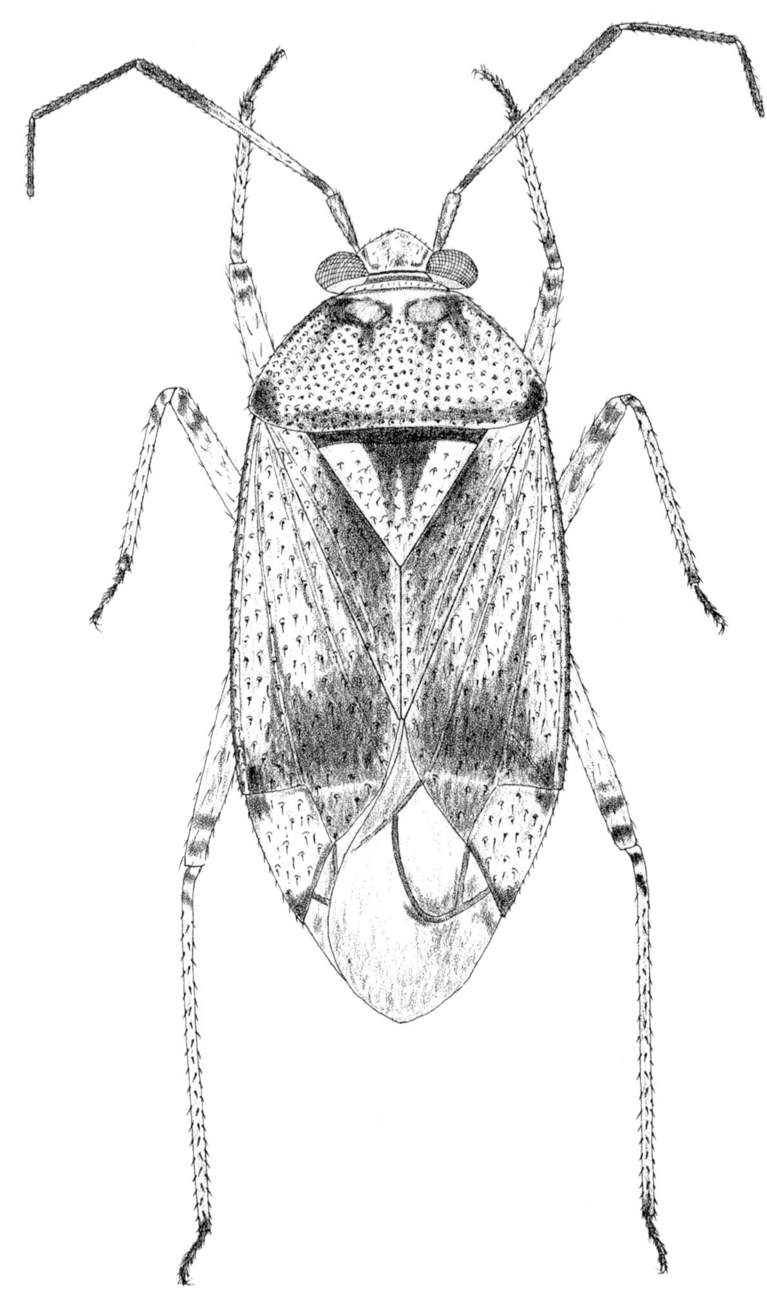

Fig. 31. *Lygus shulli*

44

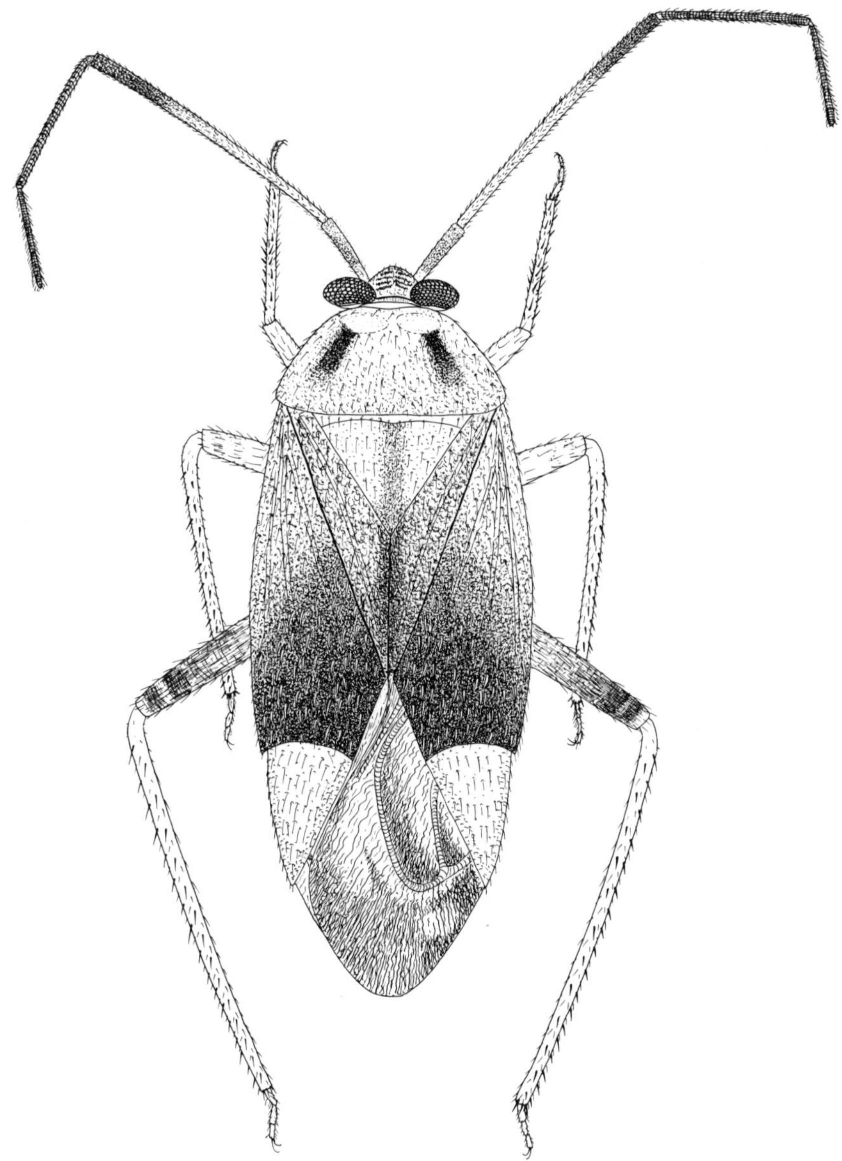

Fig. 32. *Lygocoris communis*

Table 2.   Pear plant bug on fruit crops

	B.C.	Alta.	Sask.	Man.	Ont.	Que.	N.B.	N.S.	P.E.I.	Nfld.
apple	x				x	x	x	x	x	
pear	x				x	x		x		
peach	x				x					
plum	x				x	x		x		
apricot	x				x					
sweet cherry	x				x					
sour cherry	x				x					
black cherry					x	x	x	x		
pin cherry		x	x	x	x	x	x	x	x	x
chokecherry	x	x	x	x	x	x	x	x	x	x
raspberry	x	x	x	x	x	x	x	x	x	x
blackberry	x									
thimbleberry	x				x	x	x	x		
loganberry	x									
currant	x	x	x	x	x	x	x	x	x	x
gooseberry	x	x	x	x	x	x	x	x	x	x
serviceberry	x	x	x	x	x	x	x	x	x	x
cranberry	x	x	x	x	x	x	x	x	x	x
viburnum	x		x	x	x	x	x	x		
grape	x				x					
elderberry	x	x			x	x	x			
mulberry					x					

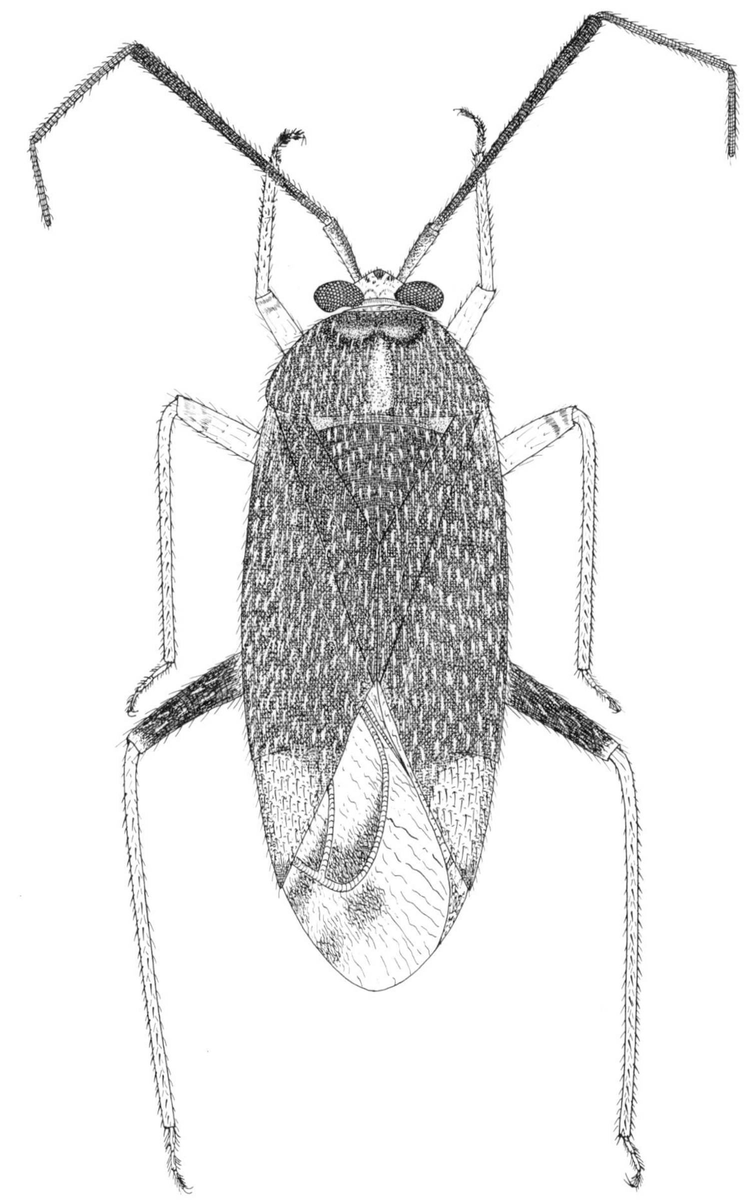

Fig. 33. *Lygocoris caryae*

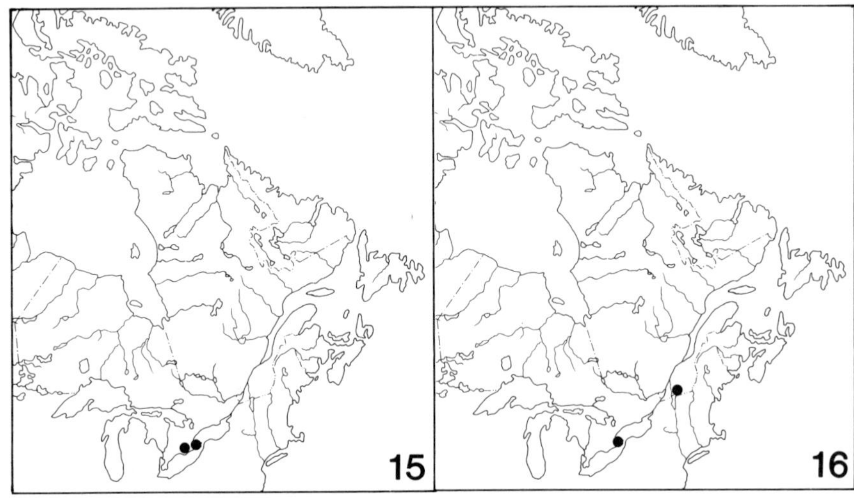

Map 15. Collection localities for *Lygocoris caryae*.

Map 16. Collection localities for *Lygocoris inconspicuus*.

Lygocoris inconspicuus (Knight)

Figs. 34,42; Map 16

Lygus (Neolygus) inconspicuus Knight, 1917*b*:612.
Neolygus inconspicuus Knight, 1941*b*:161.
Lygocoris (Neolygus) inconspicuus: Carvalho, 1959:143.

Length 4.2–4.8 mm; width 1.9–2.2 mm. Head, pronotum, scutellum, and hemelytra green; clavus and apical corium brown. Ventral surface green.

Remarks. This species is distinguished by the pattern on the hemelytra (Fig. 34), and by the claspers (Fig. 42).

Collected on wild grape in Quebec; on cultivated grape in abandoned orchards in Ontario; phytophagous.

Also collected on *Fagus grandifolia*.

Distribution. Eastern USA; Quebec, Ontario (Map 16).

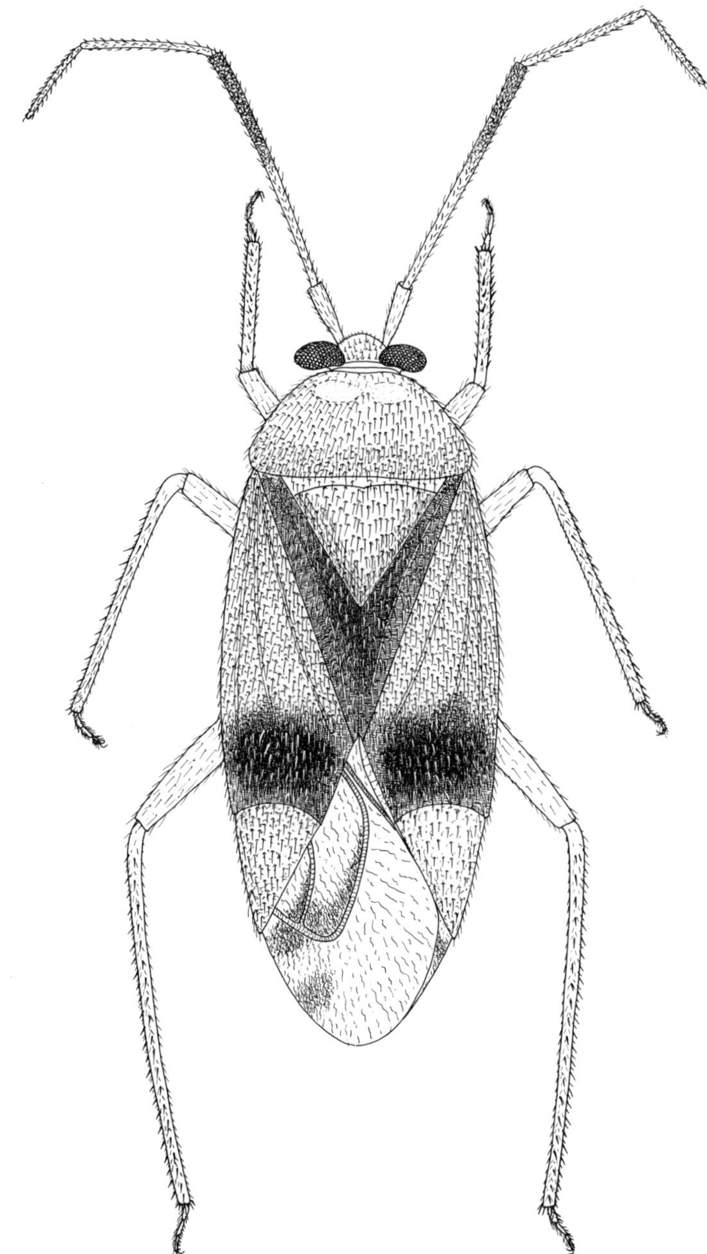

Fig. 34. *Lygocoris inconspicuus*

Lygocoris belfragii (Reuter)

Figs. 35, 43; Map 17

Lygus belfragii Reuter, 1875*b*:71.
Neolygus belfragii: Knight, 1941*b*:162.
Lygocoris (Neolygus) belfragii: Carvalho, 1959:141.

Length 5.4–6.0 mm; width 2.2–2.5 mm. Head, pronotum, and scutellum greenish yellow. Hemelytra yellowish green; triangular spot at apex of corium brown. Ventral surface yellowish green.

Remarks.   This species is distinguished by the yellowish green color, by the brown spot at the apex of corium (Fig. 35), and by the claspers (Fig. 43).

Collected on high bush-cranberry in Manitoba; on high bush-cranberry, currant, and gooseberry in Ontario; on currant and gooseberry in Quebec; on raspberry in the Maritime Provinces; phytophagous.

Also collected on *Corylus americana*.

Distribution.   Eastern USA; Maritime Provinces, Quebec, Ontario, Manitoba (Map 17).

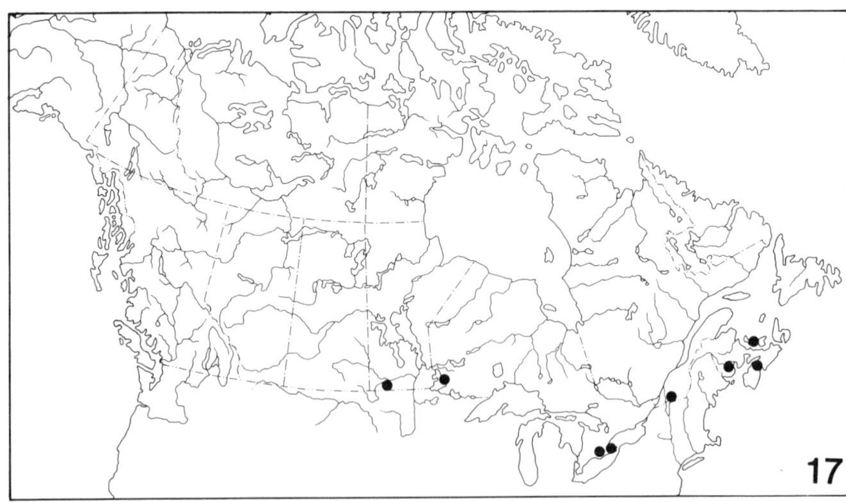

Map 17. Collection localities for *Lygocoris belfragii*.

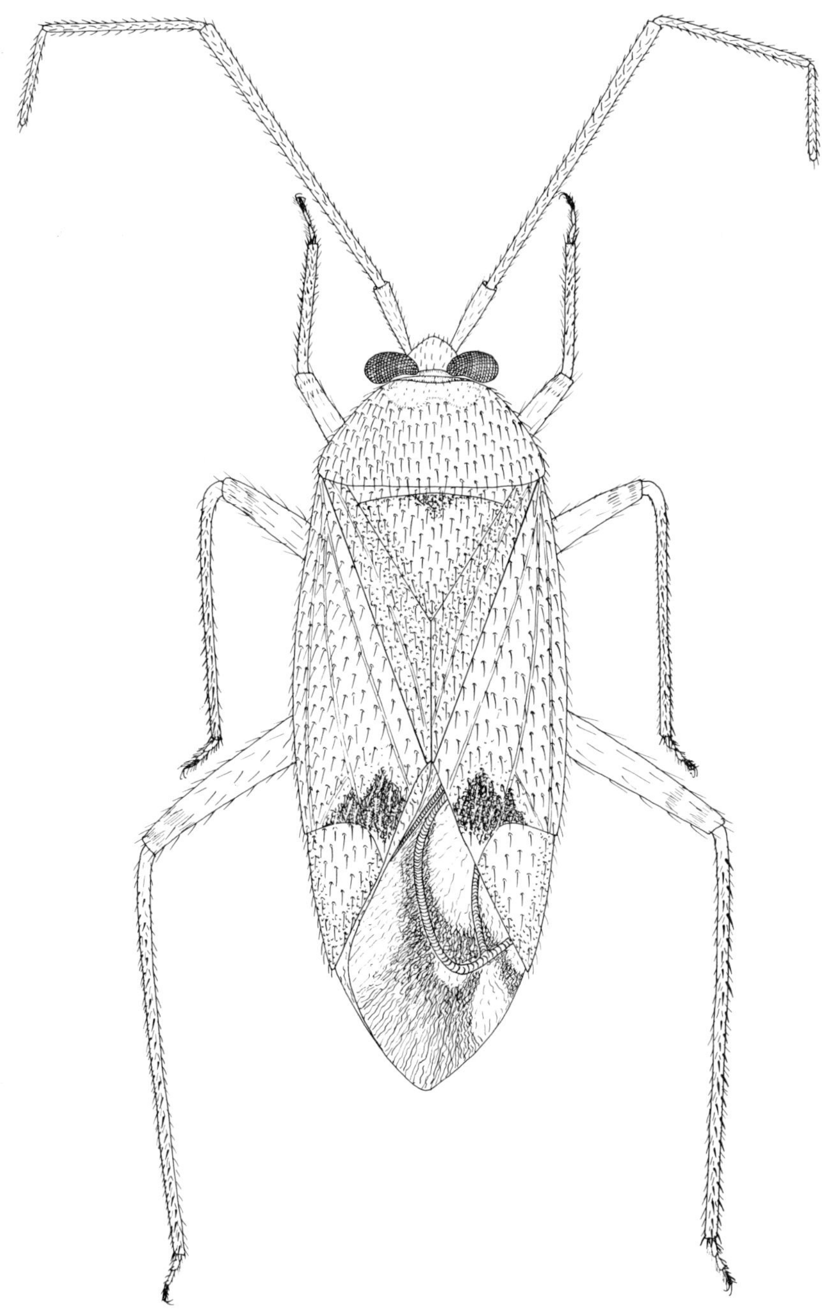

Fig. 35. *Lygocoris belfragii*

Lygocoris knighti Kelton

Figs. 36, 44; Map 18

Lygocoris (Neolygus) knighti Kelton, 1971*a*:1107.

Length 5.4–5.7 mm; width 2.2–2.4 mm. Head green, apex of clypeus black. Second antennal segment black. Pronotum green, ray behind each callus dark brown. Scutellum green. Hemelytra pale green; clavus and apical half of corium dark brown. Ventral surface light green, side of abdomen black.

Remarks. This species is distinguished by the black second antennal segment, by the dark brown rays behind the calli (Fig. 36), and by the claspers (Fig. 44).

Collected on high bush-cranberry in Manitoba and Ontario; phytophagous.

Distribution. Pennsylvania; Manitoba, Ontario (Map 18).

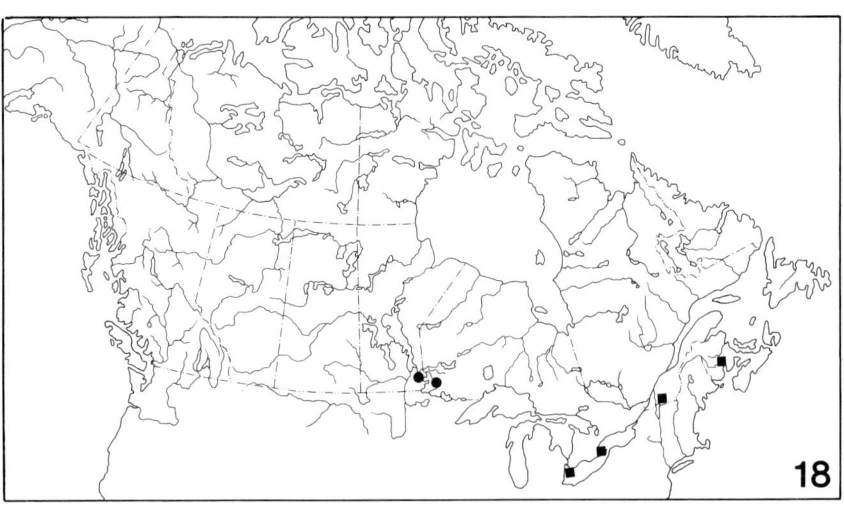

Map 18. Collection localities for *Lygocoris knighti* (●), and *Lygocoris viburni* (■).

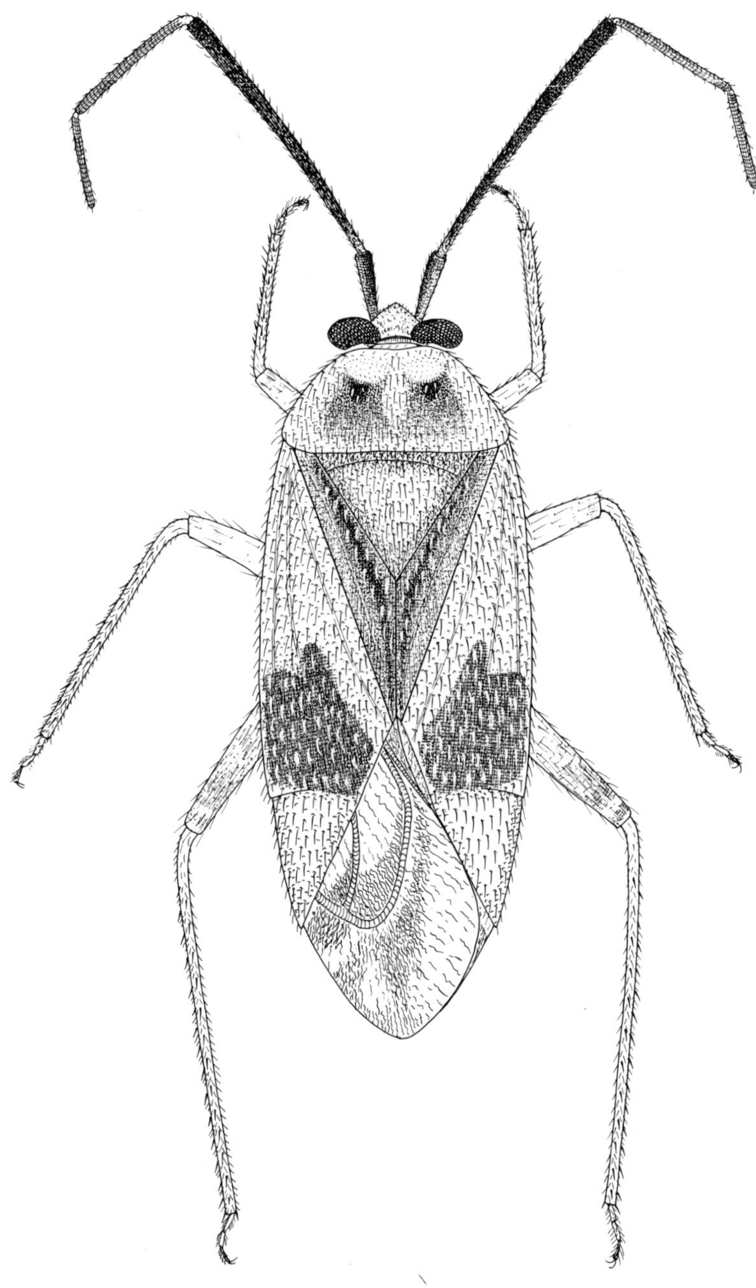

Fig. 36. *Lygocoris knighti*

Lygocoris viburni (Knight)

Figs. 37, 45; Map 18

Lygus (Neolygus) viburni Knight, 1917*b*:609.
Neolygus viburni Knight, 1941*b*:159.
Lygocoris (Neolygus) viburni: Carvalho, 1959:145.

Length 4.7–5.4 mm; width 2.1–2.4 mm. Head yellowish brown, tip of clypeus brown; frons often with transverse reddish lines. Pronotum yellowish brown. Scutellum yellowish brown, median longitudinal line reddish. Hemelytra mostly dark brown. Ventral surface yellowish green, side of abdomen dark brown.

Remarks.   This species is distinguished by the yellowish brown pronotum and scutellum, by the dark brown hemelytra (Fig. 37), and by the claspers (Fig. 45).

Collected on Canada plum, wild plum, and cultivated plum in abandoned orchards in Ontario; on Canada plum in Quebec and New Brunswick; phytophagous.

Also collected on *Viburnum lentago*.

Distribution.   Northeastern USA; Prince Edward Island, New Brunswick, Quebec, Ontario (Map 18).

Lygocoris omnivagus (Knight)

Figs. 38, 46; Map 19

Lygus (Neolygus) omnivagus Knight, 1917*b*:627.
Neolygus omnivagus Knight, 1941*b*:163.
Lygocoris (Neolygus) omnivagus: Carvalho, 1959:114.

Length 4.9–5.6 mm; width 2.1–2.4 mm. Head yellowish, tip of clypeus, lorum, and jugum brown. Pronotum yellowish, area behind callus often brown. Scutellum green. Hemelytra green; clavus and apical half of corium dark brown. Ventral surface pale green, side of abdomen dark brown.

Remarks.   This species is distinguished by the green pronotum and scutellum, by the pattern on the hemelytra (Fig. 38), and by the claspers (Fig. 46).

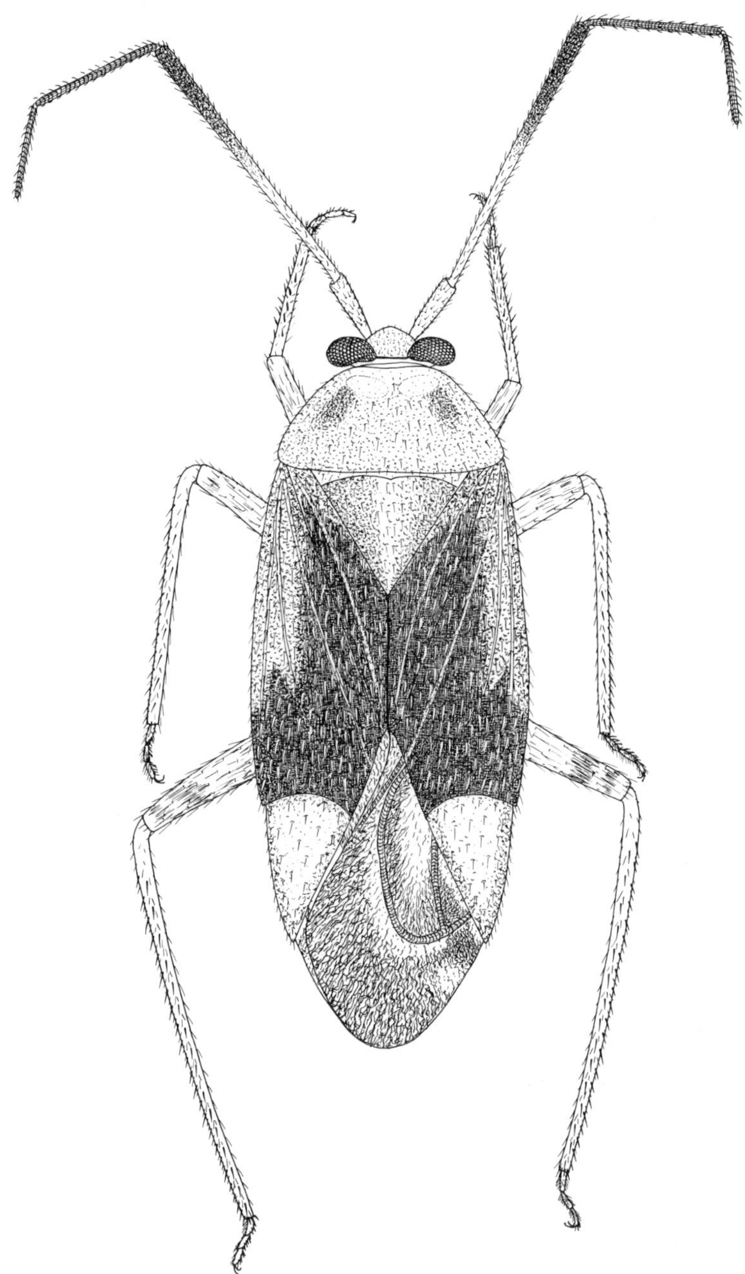

Fig. 37. *Lygocoris viburni*

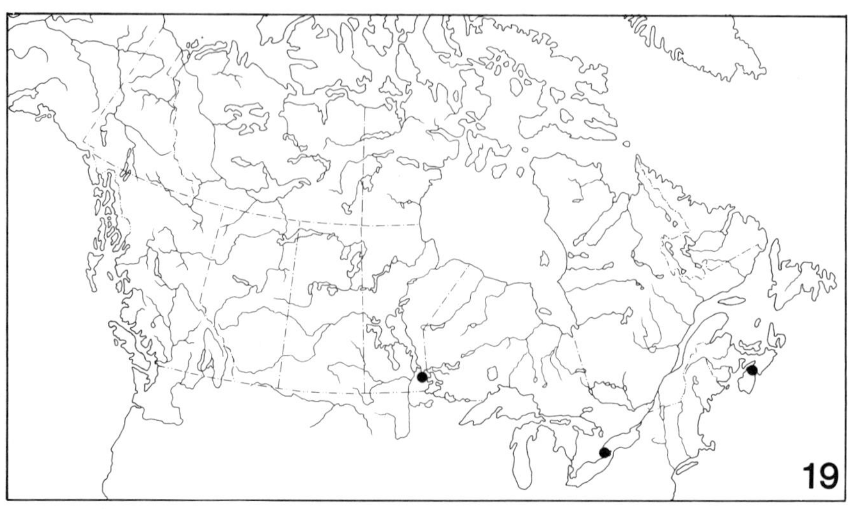

Map 19. Collection localities for *Lygocoris omnivagus*.

Collected on high bush-cranberry in Manitoba; on apple, pear, peach, apricot, sweet cherry, sour cherry, and mulberry in Ontario; on Allegheny serviceberry in Nova Scotia; phytophagous. Ross and Caesar (1921) reported the species as a pest of peach in Ontario.

Also collected on *Quercus rubra, Q. alba, Q. macrocarp, Tilia americana, Carya ovata,* and *Juglans nigra*; adults readily migrate to orchard trees and feed on the fruit, especially if the fruit trees are nearby.

Distribution.   Eastern USA; Nova Scotia, New Brunswick, Quebec, Ontario, Manitoba (Map 17).

Lygocoris quercalbae (Knight)

Figs. 39, 47; Map 20

Lygus (Neolgus) quercalbae Knight, 1917*b*:624.
Neolygus quercalbae Knight, 1941*b*:160.
Lygocoris (Neolygus) quercalbae: Carvalho, 1959:145.

Length 4.7–5.7 mm; width 2.1–2.5 mm. Head light yellowish brown marked with red. Pronotum yellowish brown marked with red, calli often brown. Scutellum green, side margins brown. Hemelytra yellowish brown; basal half of corium green. Ventral surface reddish brown.

Remarks.   The species is commonly known as the oak bug. It is distinguished by the red markings on the head and pronotum, by the brown calli (Fig. 39), and by the claspers (Fig. 47).

56

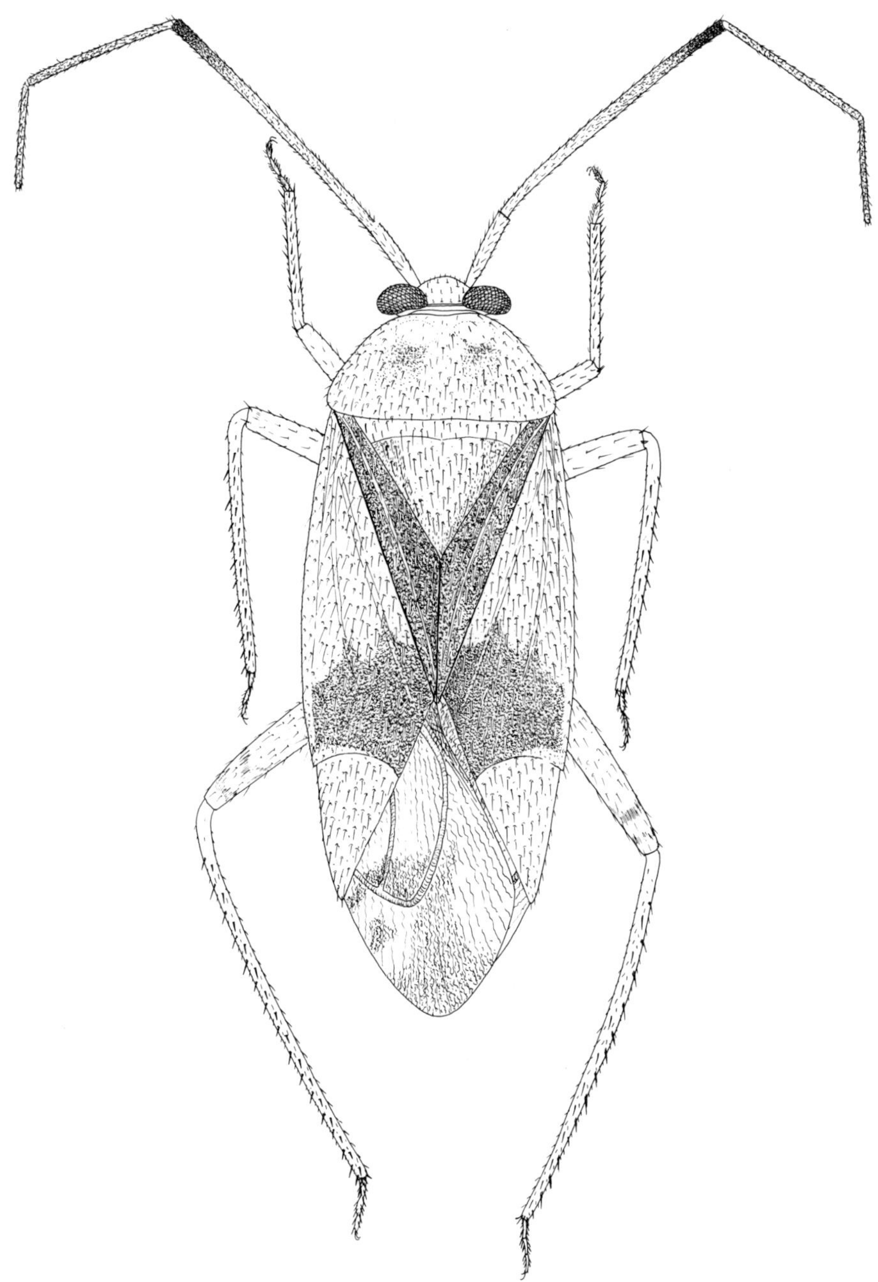

Fig. 38. *Lygocoris omnivagus*

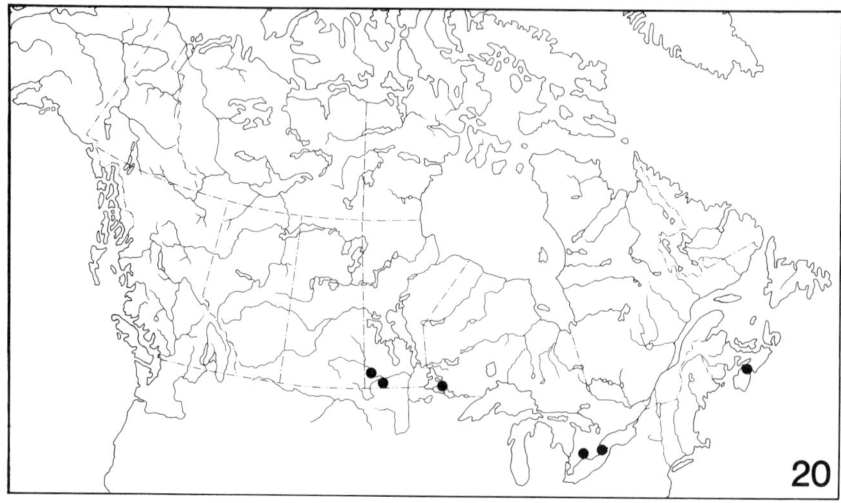

Map 20. Collection localities for *Lygocoris quercalbae*.

Collected on Allegheny serviceberry in Nova Scotia; on peach in Ontario; on saskatoon in Manitoba; phytophagous. Caesar (1920) reported the species on peach in Ontario.

Breeds on *Quercus alba* and *Q. rubra*; adults readily migrate to orchard trees and feed on the fruit, especially if the fruit trees are nearby.

Distribution.   Northeastern USA; Nova Scotia, Quebec, Ontario, Manitoba (Map 20).

Genus *Stenotus* Jakovlev

Elongate, green and black species. Head oblique; eyes large, carina between them absent. Hemelytra finely punctate, pubescence simple, dense. First tarsal segment longer than second.

One species, introduced from Europe, was collected. Overwinters in the egg stage.

Stenotus binotatus (Fabricius)

Fig. 48; Map 21

Lygaeus binotatus Fabricius, 1794:172.
Stenotus binotatus: Reuter, 1888:636.

58

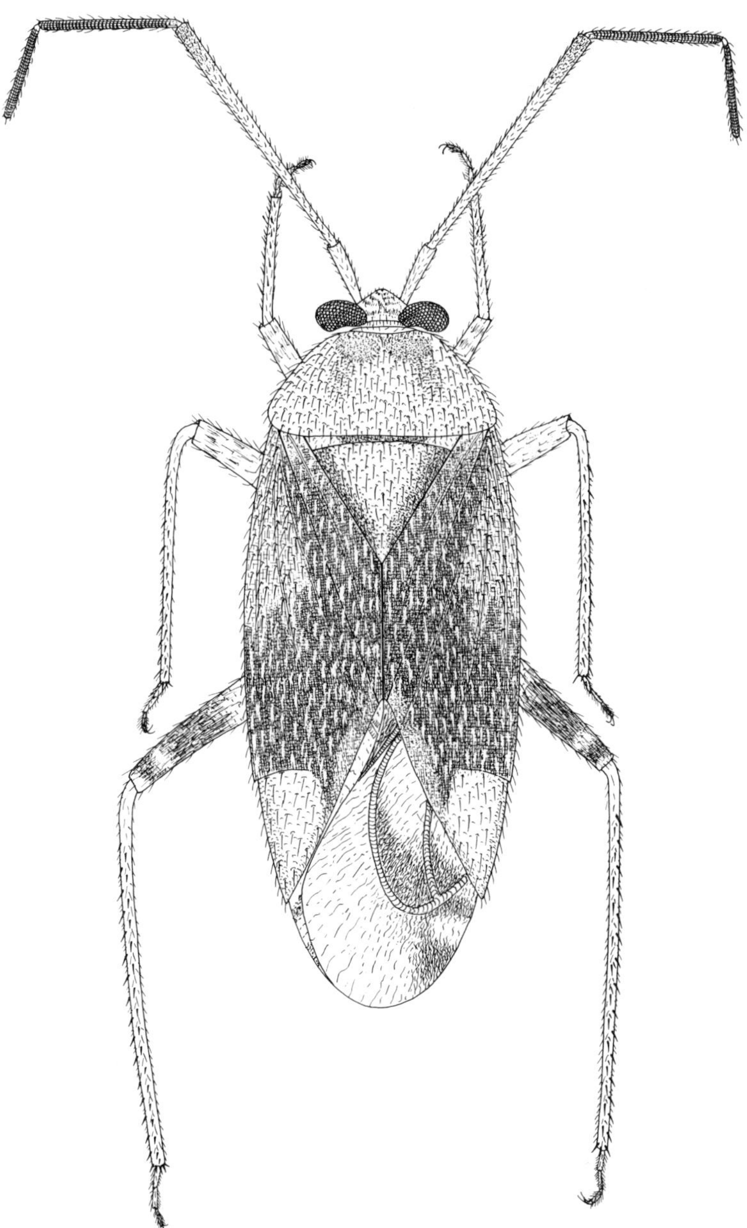

Fig. 39. *Lygocoris quercalbae*

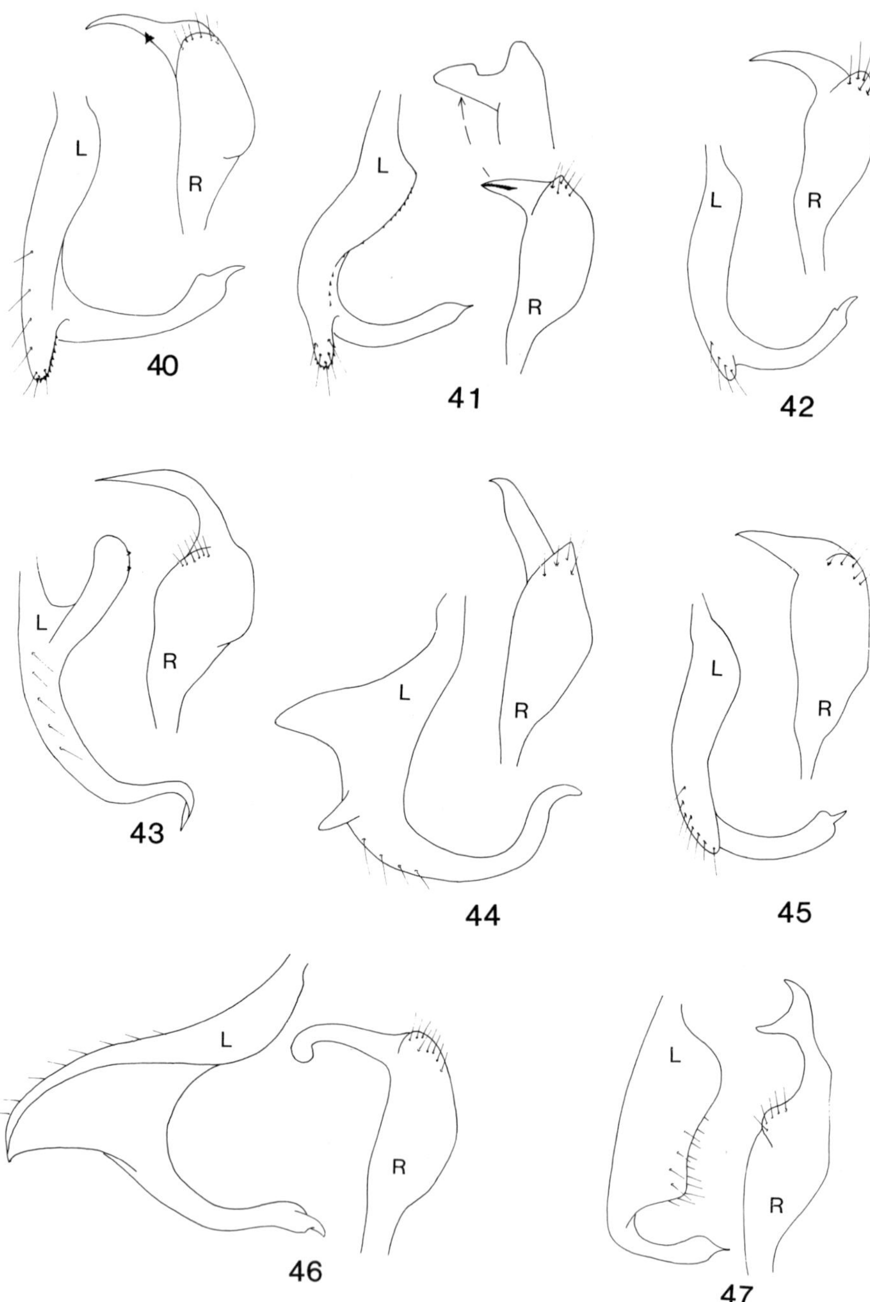

Figs. 40–47. Male claspers of *Lygocoris* spp. 40, *communis*; 41, *caryae*; 42, *inconspicuus*; 43, *belfragii*; 44, *knighti*; 45, *viburni*; 46, *omnivagus*; 47, *quercalbae*.

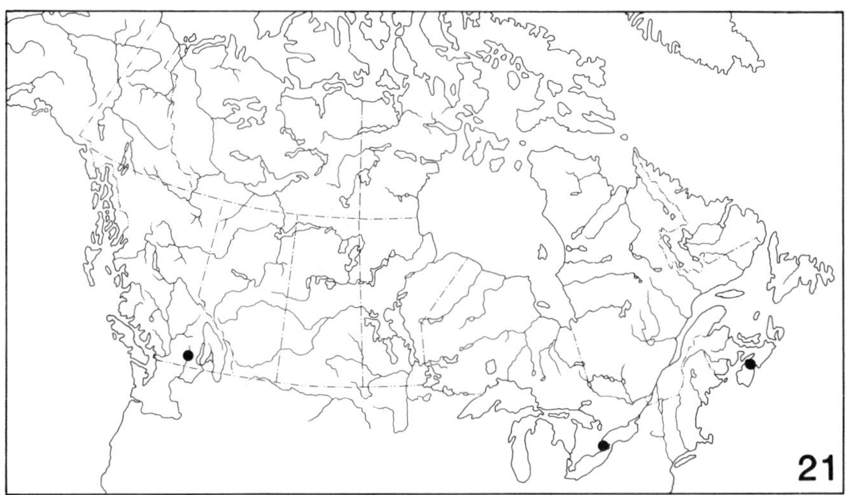

Map 21. Collection localities for *Stenotus binotatus.*

Length 5.9–6.4 mm; width 1.8–2.1 mm. Head green, clypeus and adjacent frons black. Pronotum, scutellum, and hemelytra green or yellow; stripe extending from callus to base of cuneus black. Ventral surface green.

Remarks. Osborn (1892) first listed this European species from North America. It is distinguished by the two black stripes on the hemelytra (Fig. 48).

Collected on apple and pear in Nova Scotia; on apple, pear, plum, peach, and sweet cherry in Ontario and British Columbia.

Breeds on orchard grasses, but when the grasses are cut or during the dry season the adults migrate to fruit trees and feed on the foliage or fruit.

The nymphs appear in May or earlier and the adults in June. By the end of July most of the adults die out.

Distribution. Holarctic; transcontinental in USA; Nova Scotia, Quebec, Ontario, Manitoba, British Columbia (Map 21).

Genus *Calocoris* Fieber

Robust, green species. Head oblique; carina between eyes absent. Pronotum finely rugose. Hemelytra finely punctate; pubescence of two types, sericeous hairs intermixed with simple black hairs. Genital segment with stout tubercle near base of left clasper.

One species, introduced from Europe, was collected. Overwinters in the egg stage.

Calocoris norvegicus (Gmelin)

Fig. 49; Map 22

Cimex norvegicus Gmelin, 1788:2176.
Calocoris bipunctatus Provancher, 1886:114.
Calocoris norvegicus: Reuter, 1888:232.

Length 7.1–7.5 mm; width 2.6–3.0 mm. Head yellowish green. First antennal segment green, often marked with black; second segment greenish brown. Pronotum green, spot behind each callus black. Scutellum green. Hemelytra green; in older males clavus and corium often tinged with reddish brown. Legs green, femora spotted with black.

Remarks. Provancher (1886) first reported this European species from Quebec. It is distinguished by the robust size, by the two black spots on the pronotum, and by the green or reddish brown hemelytra (Fig. 49). The tubercle on the genital segment near the base of each clasper is present.

Collected on strawberry in Prince Edward Island, Nova Scotia, New Brunswick, and British Columbia; phytophagous. Pickett (1943) reported it as a pest of strawberry in Nova Scotia.

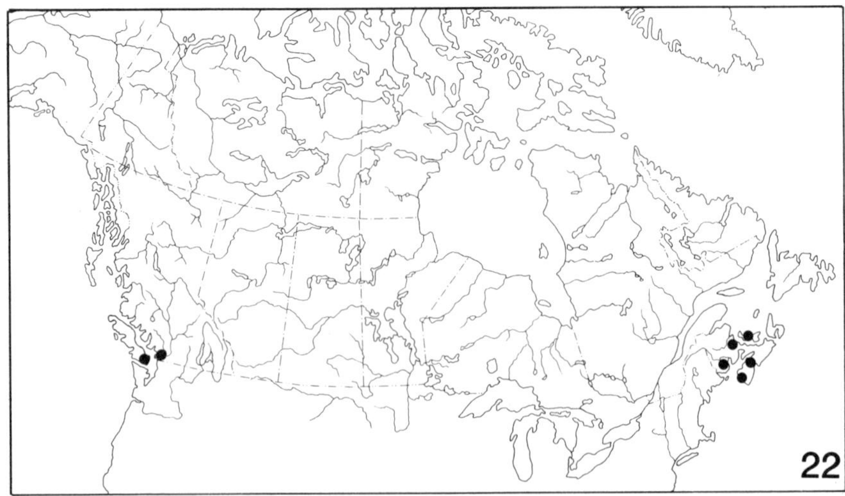

Map 22. Collection localities for *Calocoris norvegicus*.

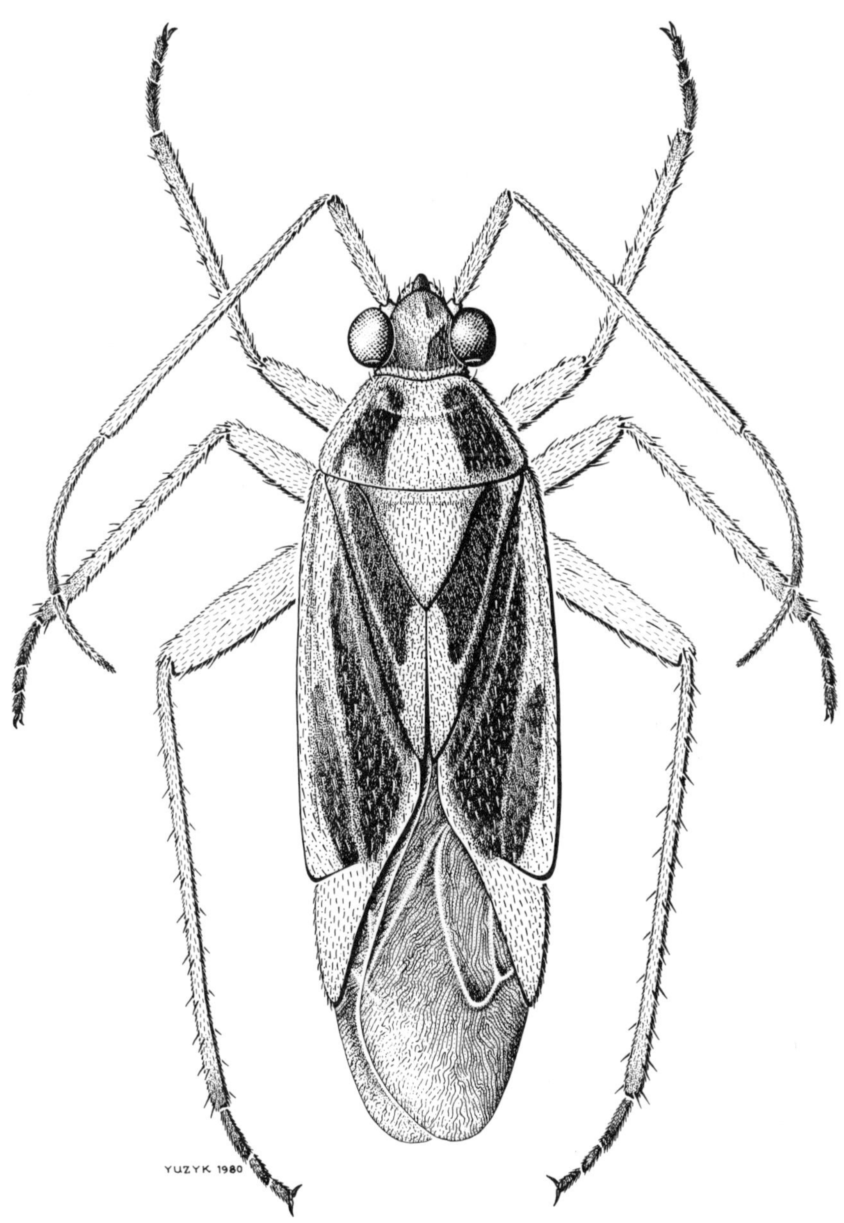

Fig. 48. *Stenotus binotatus*

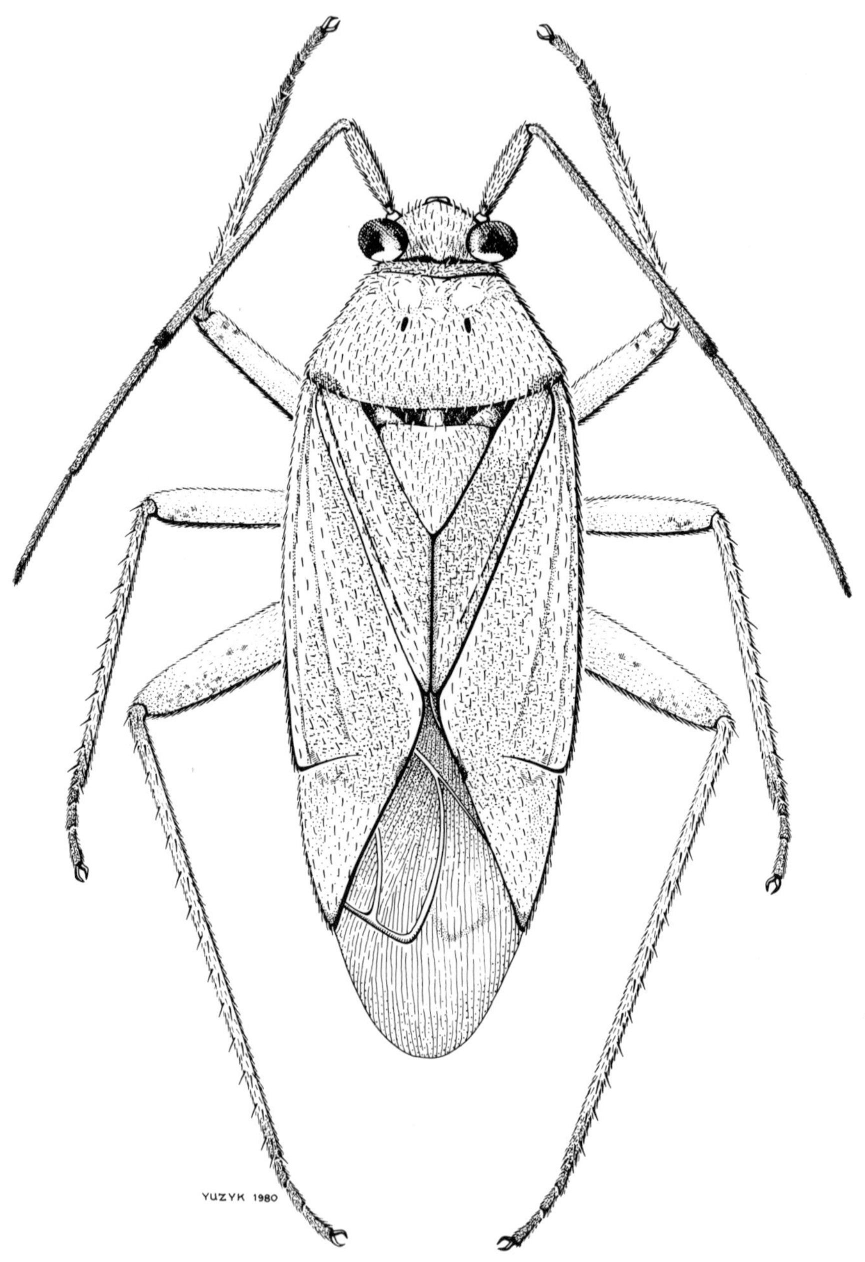

Fig. 49. *Calocoris norvegicus*

Also collected on many other plants.

The nymphs appear in June and the adults in July. By mid-August most of the adults die out.

Distribution.   Holarctic; northeastern USA; Maritime Provinces, British Columbia (Map 22).

Genus *Phytocoris* Fallén

Elongate, parallel species. Head oblique, short, lora inflated; eyes large and prominent, carina between them absent. Pronotum impunctate. Pubescence of two types, appressed sericeous hairs intermixed with simple slanting hairs. Legs long and slender.

The genus is large and contains many species that are similar in appearance, and the females are difficult to identify. The males may be identified with certainty by the shape of the claspers, and the females by association with the males.

Fifteen species were collected, one introduced from Europe. Overwinter in the egg stage. The nymphs appear in June and the adults in July. By the end of August most of the adults die out. The species are predaceous on all soft-bodied arthropods found on the host plants.

Key to species of *Phytocoris*

1. Wing membrane speckled with dark spots, or with pale spots (Figs. 50–55) 2
   Wing membrane marbled (Figs. 56,57) 8
2. First antennal segment greatly thickened (Fig. 50); claspers (Fig. 56) . . . . *lasiomerus* **Reuter** (p. 66)
   First antennal segment slender (Fig. 51) 3
3. Yellow species; head, pronotum, and hemelytra speckled with red; scutellum inflated, with red spot each side near apex (Fig. 51) . . . . *interspersus* **Uhler** (p. 67)
   Brown or gray species; scutellum not inflated 4
4. Clypeus, jugum, and lorum yellow, without dark markings; claspers (Fig. 59) . . . . *sulcatus* **Knight** (p. 70)
   Clypeus, jugum, and lorum with dark markings 5
5. First antennal segment reddish brown with few small, pale spots; femora mostly reddish brown (Fig. 52); claspers (Fig. 60) . . . . *corticevivens* **Knight** (p. 70)
   First antennal segments with large pale areas; femora mostly pale with large connected brown areas (Fig. 53) 6
6. Second antennal segment with pale band at base only (Fig. 53); left clasper (Fig. 61) . . . . *gracillatus* **Knight** (p. 71)
   Second antennal segment with pale band at base and middle (Figs. 54,55) 7

7. Tubercle on genital segment near base of left clasper present (Fig. 62) ...
.... *conspurcatus* **Knight** (p. 74)
   Tubercle on genital segment near base of left clasper absent; claspers (Fig.
   63) .... *dimidiatus* **(Kirschbaum)** (p. 76)
8. Head pale green, lorum black; pronotum black, basal margin pale green;
   hemelytra mostly green (Fig. 56); right clasper (Fig. 64) ....
   .... *nigricollis* **Knight** (p. 76)
   Head gray or reddish brown; pronotum and hemelytra mostly brown ...
   9
9. Rostrum shorter than 2.8 mm 10
   Rostrum 2.8 mm or longer 12
10. Rostrum shorter than 2.4 mm, scarcely extending beyond hind coxae; right
    clasper (Fig. 65) .... *husseyi* **Knight** (p. 77)
    Rostrum longer than 2.5 mm, extending far beyond hind coxae 11
11. Left clasper with large knobbed process near base; right clasper curved with
    short knob near base (Fig. 66) .... *erectus* **Van Duzee** (p. 80)
    Left clasper with short pointed process near base; right clasper straight and
    forked (Fig. 67) .... *canadensis* **Van Duzee** (p. 81)
12. Second antennal segment mostly pale, black at apex and near base (Fig. 57);
    process at base of right clasper rounded (Fig. 68) ....
    .... *salicis* **Knight** (p. 81)
    Second antennal segment black, pale only at base 13
13. Rostrum 2.9 mm or longer; process at base of right clasper triangular
    (Fig. 69) .... *onustus* **Van Duzee** (p. 83)
    Rostrum shorter than 2.9 mm 14
14. Process at base of right clasper short (Fig. 70) .. *neglectus* **Knight** (p. 85)
    Process at base of right clasper slender (Fig. 71) ....
    .... *cortitectus* **Knight** (p. 86)

Phytocoris lasiomerus Reuter

Figs. 50, 58; Map 23

Phytocoris scrupeus Provancher, nec Say, 1886:108.
Phytocoris annulicornis Osborn, 1892:123 (n. preoc.).
Phytocoris lasiomerus Reuter, 1909:34.

Length 7.3–8.0 mm; width 2.1–2.2 mm. Head yellowish brown marked with red. First antennal segment thickened with long, black bristles; second segment pale yellow, apical one fourth black. Pronotum, scutellum, and hemelytra yellowish brown, often with reddish tinge; simple hairs black and light brown.

Remarks. This species is distinguished by the thickened first antennal segment (Fig. 50) and by the claspers (Fig. 58). A tubercle on the genital segment near the base of each clasper is present.

Collected on apple in Ontario and Quebec; on apple, pear, peach, and cherry in British Columbia; predaceous on aphids. Braimah et al. (1981) observed the species on apple in Quebec.

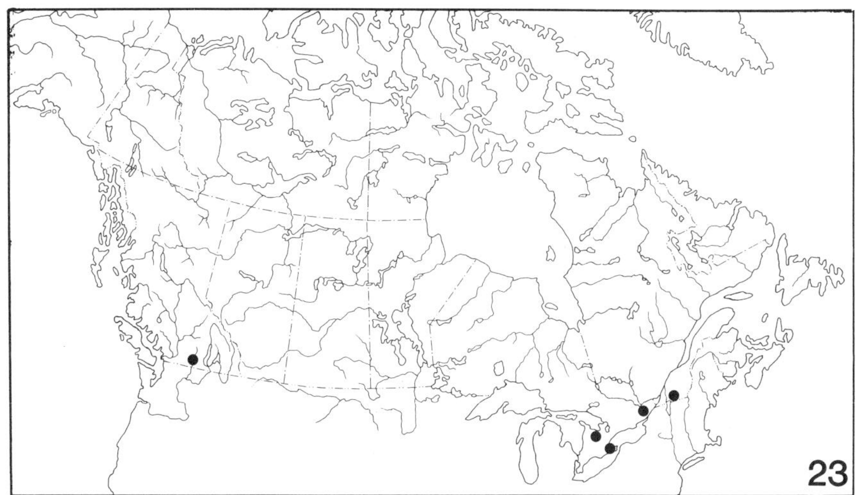

Map 23. Collection localities for *Phytocoris lasiomerus*.

Also collected on many herbaceous plants infested with aphids.

Distribution.   Northern USA; British Columbia, Prairie Provinces, Ontario, Quebec (Map 23).

Phytocoris interspersus Uhler

Fig. 51; Map 24

Phytocoris interspersus Uhler, 1895:32.

Length 6.4–6.8 mm; width 2.1–2.2 mm. Head yellow often marked with red. First antennal segment slender with few red spots, second segment yellow. Pronotum yellow, older specimens marked with red. Scutellum yellow, inflated, spot on each side near apex red. Hemelytra yellow irrorate with red; simple pubescence pale yellow. Legs yellow, femora often marked with red.

Remarks.   This species is distinguished by the overall yellow color often irrorate with red, and by the inflated scutellum with red dots near apex (Fig. 51). There are no tubercles on the genital segment.

Collected on peach and sweet cherry in British Columbia; predaceous on aphids.

Also collected on *Shepherdia canadensis* and *Quercus* spp. infested with aphids.

Distribution.   Western USA; British Columbia (Map 24).

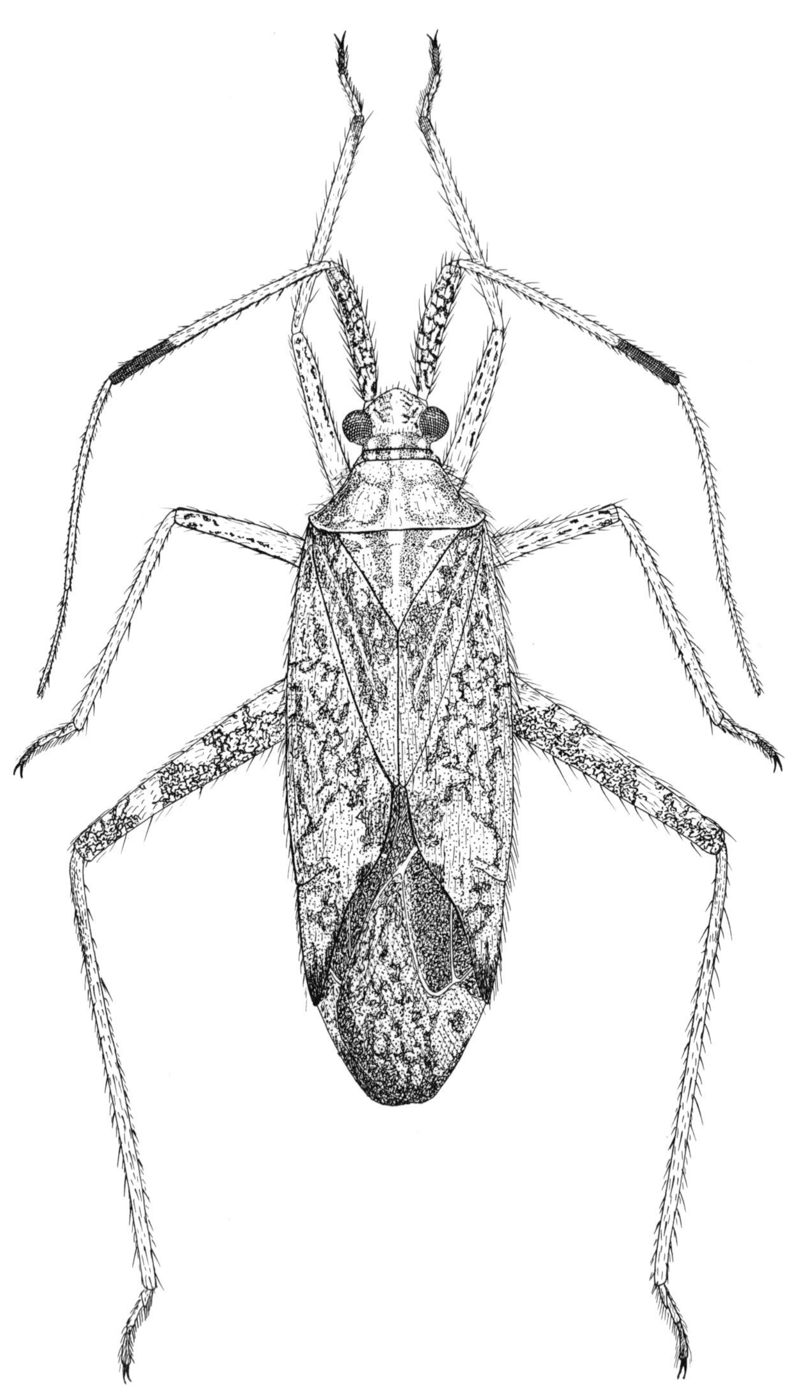

Fig. 50. *Phytocoris lasiomerus*

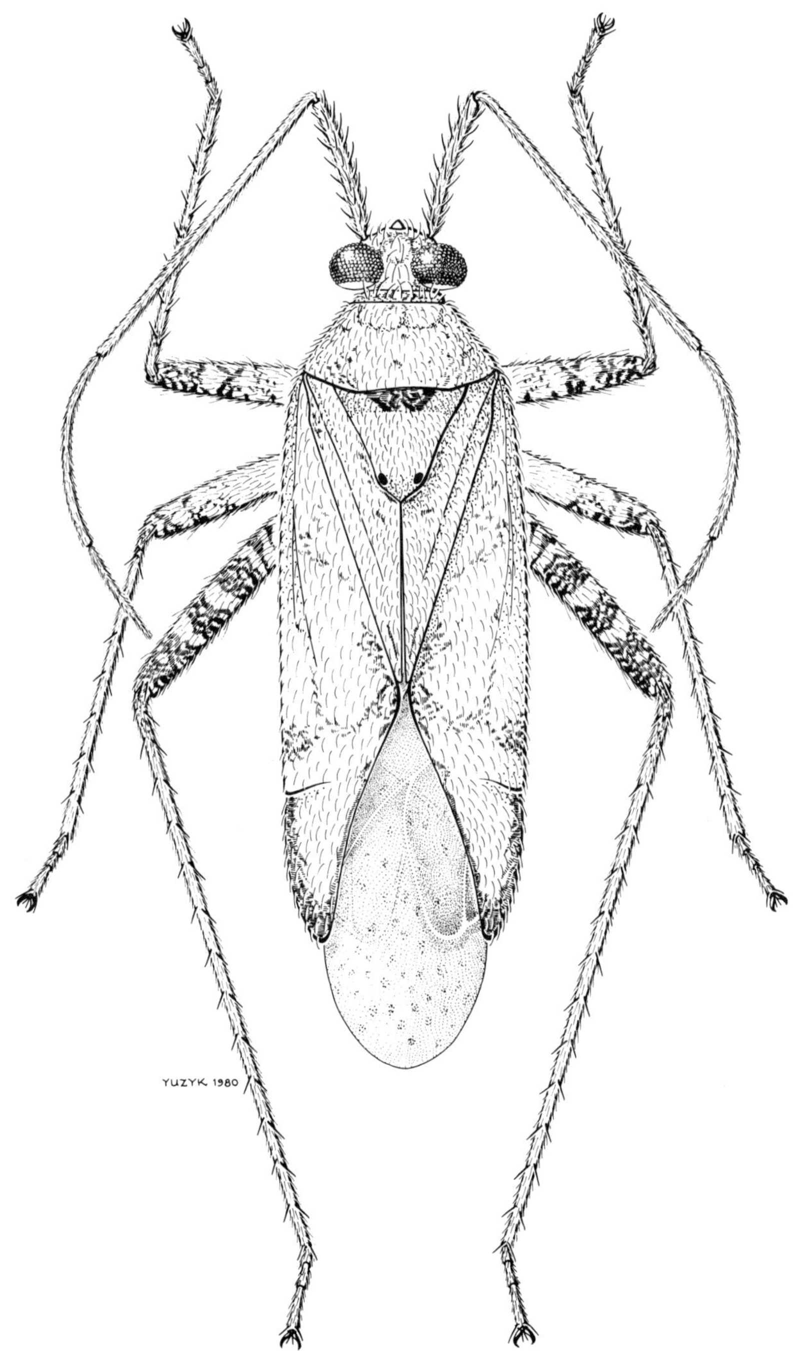

Fig. 51. *Phytocoris interspersus*

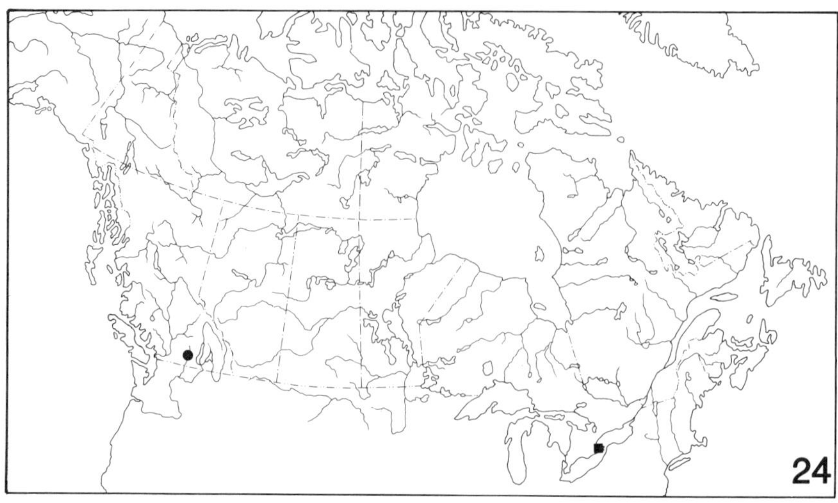

Map 24. Collection localities for *Phytocoris interspersus* (●), and *Phytocoris sulcatus* (■).

Phytocoris sulcatus Knight

Fig. 59; Map 24

Phytocoris sulcatus Knight, 1920:64.

Length 5.6–6.1 mm; width 1.8–1.9 mm. Head yellow below ventral margins of eyes, brown above. Rostrum 2.8–3.1 mm long. First antennal segment brown marked with yellow spots; second segment brown, base and middle yellow. Pronotum brown, ray behind callus and usually four spots along subbasal margin black. Scutellum brown, longitudinal median line yellow margined with black. Hemelytra gray marked with black; apex of corium with large gray spot; simple hairs black. Legs pale marked with black.

Remarks. This species is distinguished by the pale lower half of the head, and by the claspers (Fig. 59). The tubercle on the genital segment near the base of each clasper is present.

Collected on apple, pear, peach, apricot, sweet cherry, and mulberry in Ontario; predaceous on aphids, mites, and other small arthropods.

Distribution. Eastern and north central USA; Ontario (Map 24).

Phytocoris corticevivens Knight

Figs. 52, 60; Map 25

Phytocoris corticevivens Knight, 1920:63.

Length 6.6–6.8 mm; width 2.3–2.5 mm. Head brown, bar next to eye on vertex yellow. First antennal segment long and slender, reddish brown, with several pale spots; second segment reddish brown, base and narrow area at middle pale. Pronotum dark brown. Scutellum dark brown, median longitudinal line pale. Hemelytra dark brown; rounded spot at apex of corium pale; simple hairs black. Femora mostly brown.

Remarks. This species is distinguished by the long and slender first antennal segment, by the overall dark brown color (Fig. 52), and by the claspers (Fig. 60). A tubercle on the genital segment near the base of each clasper is present.

Collected on the trunks of apple and pear in Nova Scotia, Quebec, and Ontario, usually hiding in the crevices of the bark; predaceous on all small arthropods.

Distribution. Texas, north central and northeastern USA; Ontario, Quebec, Nova Scotia (Map 25).

Phytocoris gracillatus Knight

Figs. 53, 61; Map 25

Phytocoris gracillatus Knight, 1968:229.

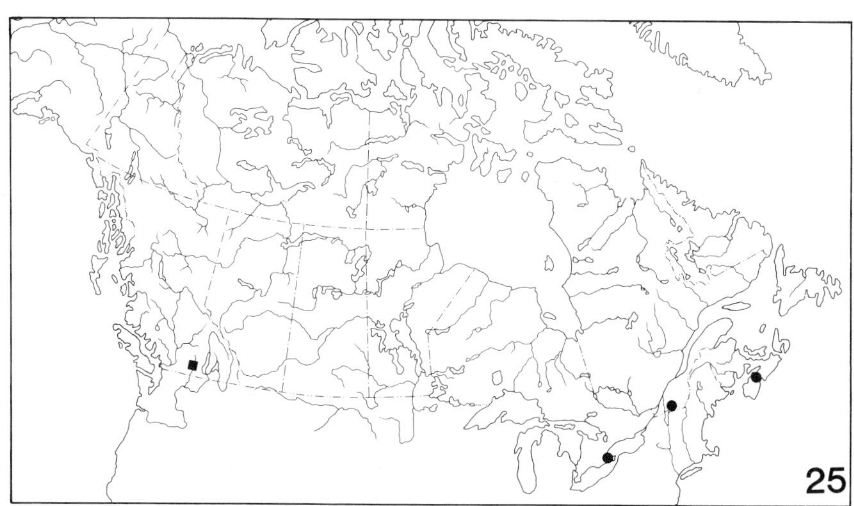

Map 25. Collection localities for *Phytocoris corticevivens* (●), and *Phytocoris gracillatus* (■).

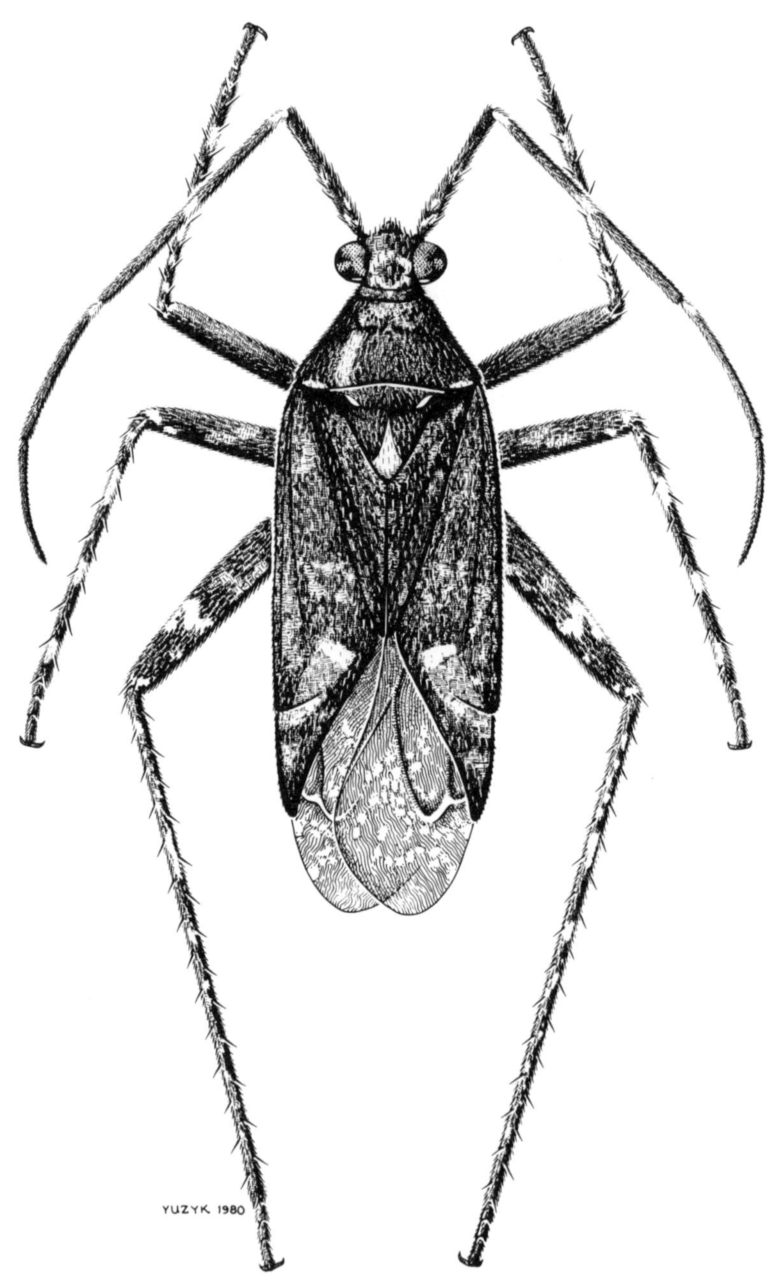

Fig. 52. *Phytocoris corticevivens*

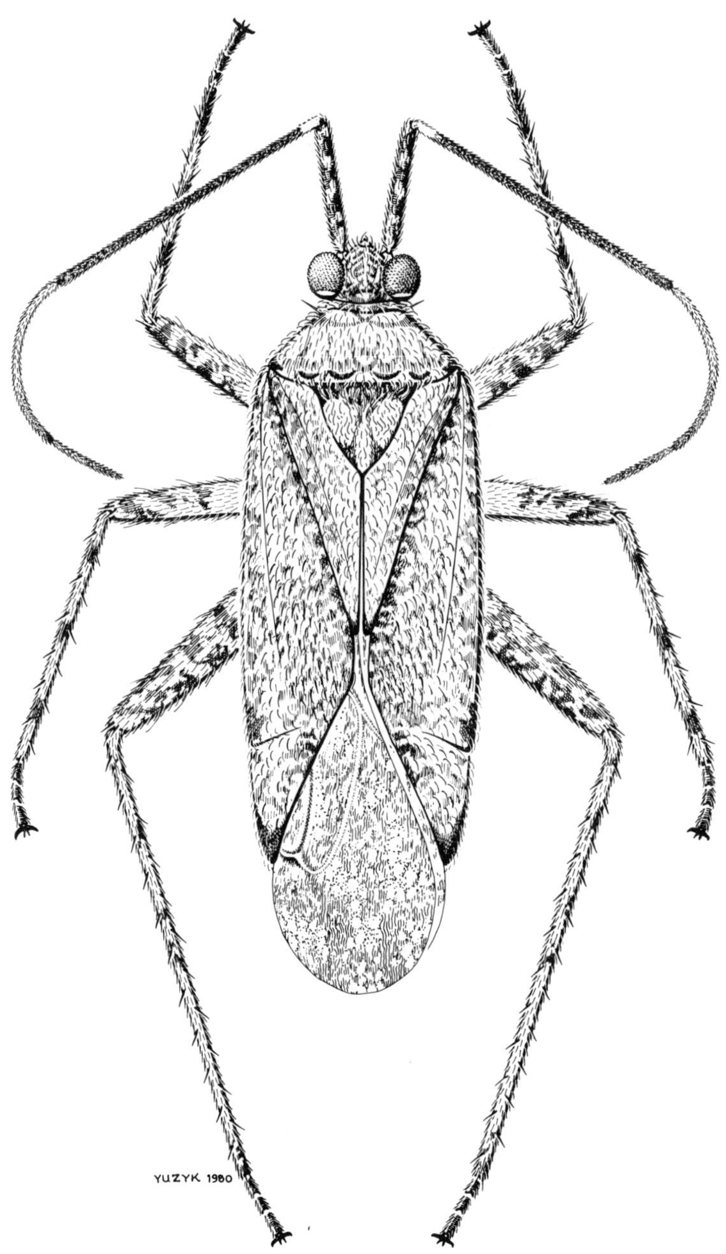

Fig. 53. *Phytocoris gracillatus*

Length 6.5–7.0 mm; width 1.9–2.2 mm. Head gray marked with black. First antennal segment pale on ventral surface, dorsal surface mottled with black; second segment dark brown with pale band at base. Pronotum gray mottled with black. Scutellum gray, lateral margins black. Hemelytra gray mottled with black. Legs pale marked with black.

Remarks. This species is distinguished by the overall mottled gray appearance (Fig. 53) and by the left clasper (Fig. 61). There are no tubercles on the genital segment.

Collected on apple in British Columbia; predaceous on aphids.

Also collected on *Rosa nutcana* and *Artemisia tridentata*.

Distribution. Western USA; British Columbia (Map 25).

Phytocoris conspurcatus Knight

Figs. 54, 62; Map 26

Phytocoris conspurcatus Knight, 1920:61.

Length 5.7–5.9 mm; width 2.1–2.3 mm. Head gray marked with brown and reddish brown. First antennal segment gray mottled with black; second segment dark brown with white band at base and middle. Pronotum gray, subbasal margin often black. Hemelytra gray marked with brown; sericeous pubescence white and black, simple hairs black.

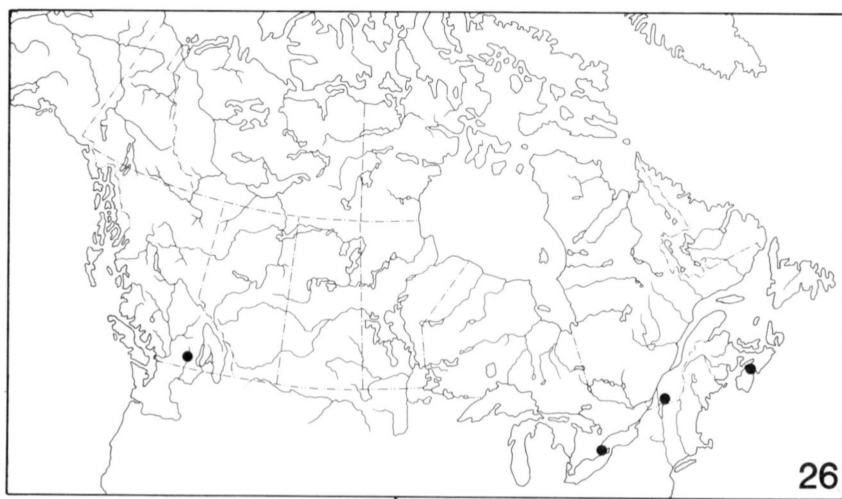

Map 26. Collection localities for *Phytocoris conspurcatus*.

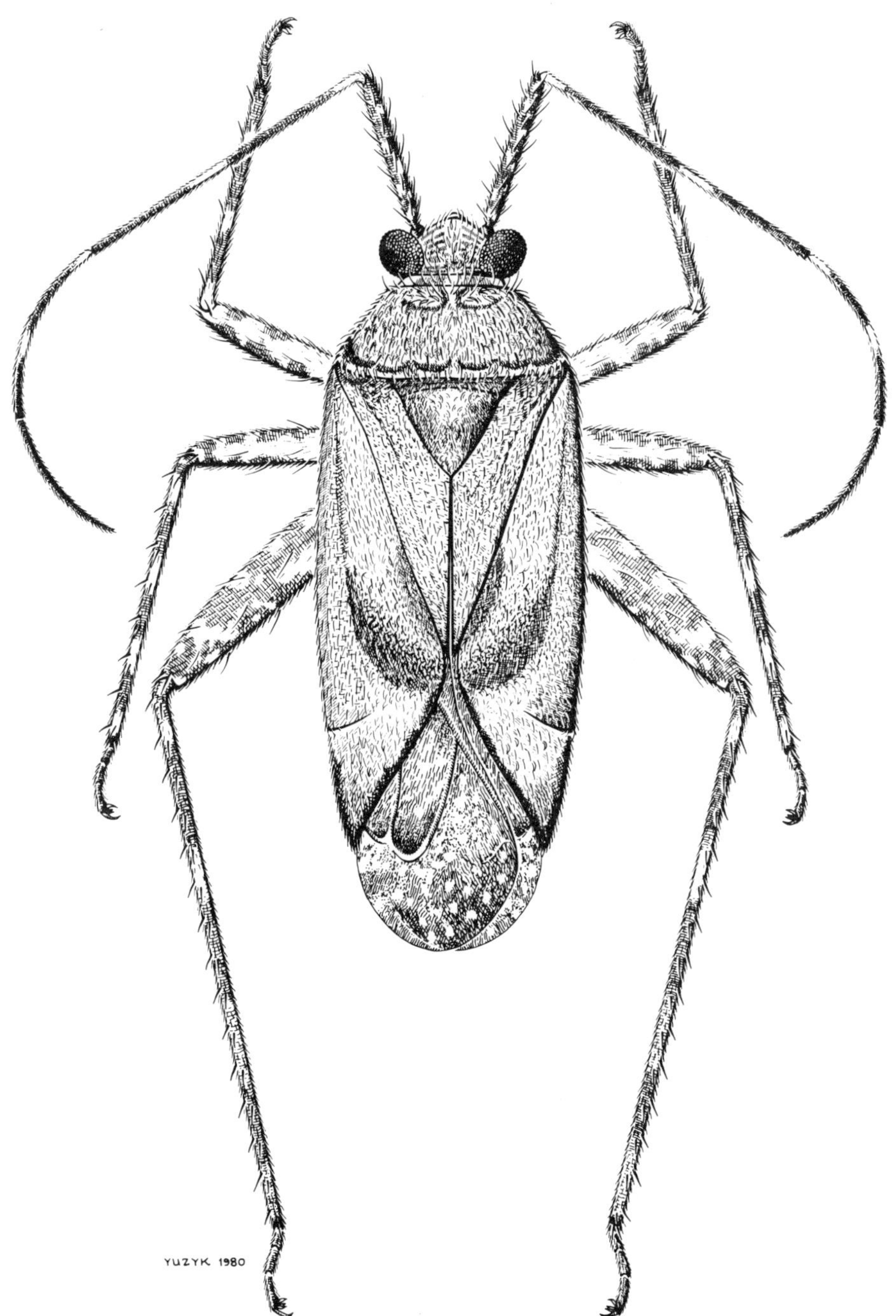

Fig. 54. *Phytocoris conspurcatus*

Remarks. This species is distinguished by the black and white sericeous hairs, by the banded second antennal segment (Fig. 54), and by the claspers (Fig. 62). A tubercle on the genital segment near the base of each clasper is present.

Collected on apple and pear in Nova Scotia and Quebec; on apple, pear, and mulberry in Ontario; on apple, peach, and sweet cherry in British Columbia; predaceous on aphids, mites, and other small arthropods. Patterson and Neary (1952) reported the species on apple in Nova Scotia. Braimah et al. (1981) observed the species on apple in Quebec. The adults often hide in the crevices on the bark.

Distribution. Eastern USA; Nova Scotia, Quebec, Ontario, Prairie Provinces, British Columbia (Map 26).

Phytocoris dimidiatus (Kirschbaum)

Figs. 55, 63; Map 27

Phytocoris dimidiatus Kirschbaum, 1855:199.

Length 6.3–6.9 mm; width 2.2–2.4 mm. Head gray marked with brown. First antennal segment pale on ventral surface, dorsal surface mottled with brown; second segment dark brown with white band at base and middle. Pronotum gray, lateral and subbasal margins black. Hemelytra brown; apex of corium with triangular pale area; sericeous pubescence silvery, simple pubescence black.

Remarks. Knight (1923*b*) first reported this European species from Nova Scotia. It is distinguished by the pale triangular area on the apical corium, by the banded second antennal segment (Fig. 55), and by the claspers (Fig. 63). There are no tubercles on the genital segment.

Collected on apple, pear, and plum in Nova Scotia; predaceous on aphids.

Distribution. Holarctic; Nova Scotia (Map 27).

Phytocoris nigricollis Knight

Figs. 56, 64; Map 28

Phytocoris nigricollis Knight, 1923*b*:636.

Length 5.1–5.3 mm; width 1.6–1.9 mm. Head pale green, lorum black. First antennal segment pale on ventral surface, dorsal surface mottled with brown; second segment dark brown with white band at base. Pronotum black, basal margin pale green, diagonal bar each side of

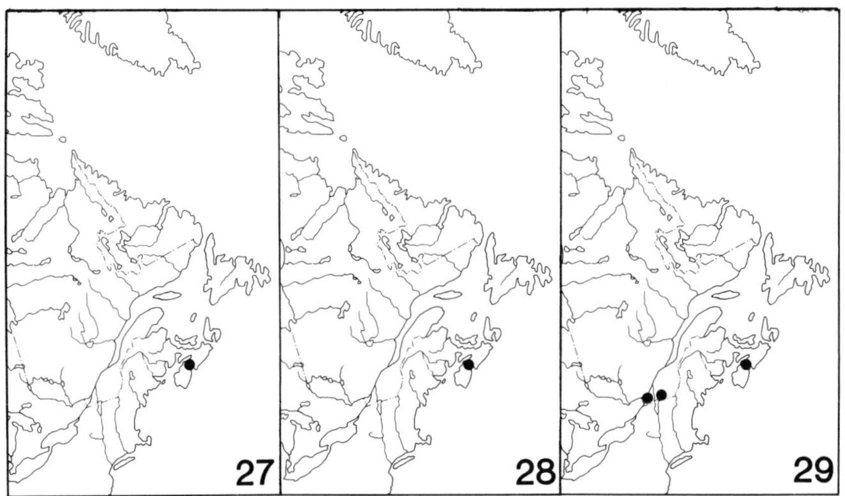

Map 27. Collection locality for *Phytocoris dimidiatus*.

Map 28. Collection locality for *Phytocoris nigricollis*.

Map 29. Collection localities for *Phytocoris husseyi*.

median line at middle black. Hemelytra pale green with black markings; pubescence silvery on pale areas, black on black areas.

Remarks. This is the only pale green species encountered. It is distinguished by the pale green head, the black pronotum, the green hemelytra (Fig. 56), and the right clasper (Fig. 64). A short tubercle on the genital segment near the base of each clasper is present.

Collected on *Malus pumila* in Nova Scotia; predaceous on aphids and mites.

Distribution. New Hampshire, North Carolina; Nova Scotia (Map 28).

Phytocoris husseyi Knight

Fig. 65; Map 29

Phytocoris husseyi Knight, 1923*b*:639.

Length 5.2–5.9 mm; width 1.9–2.1 mm. Head gray marked with dark brown. Rostrum 2.2–2.3 mm long. First antennal segment black with several white spots; second segment brown with pale band at base. Pronotum gray, lateral and subbasal margins black. Scutellum gray, diagonal bar each side of middle black. Hemelytra gray mottled with black; sericeous pubescence occurs in clumps.

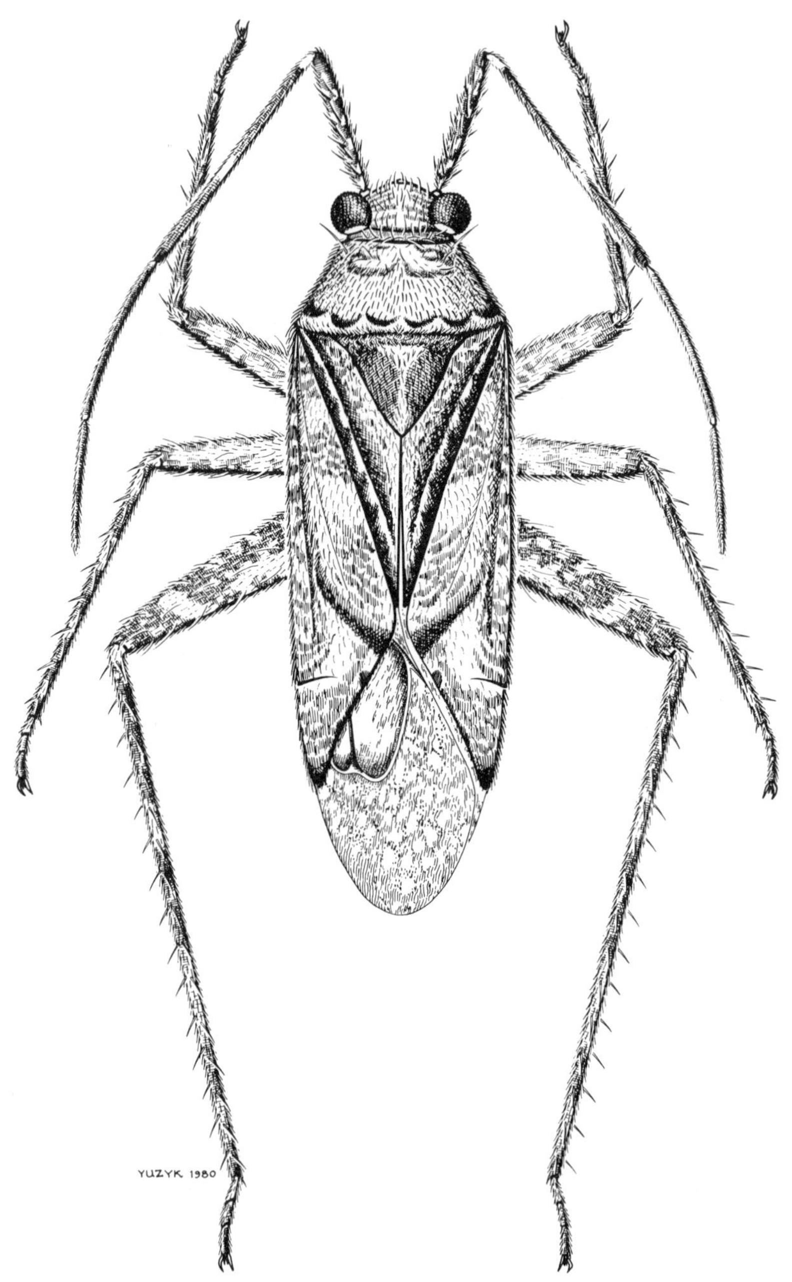

Fig. 55. *Phytocoris dimidiatus*

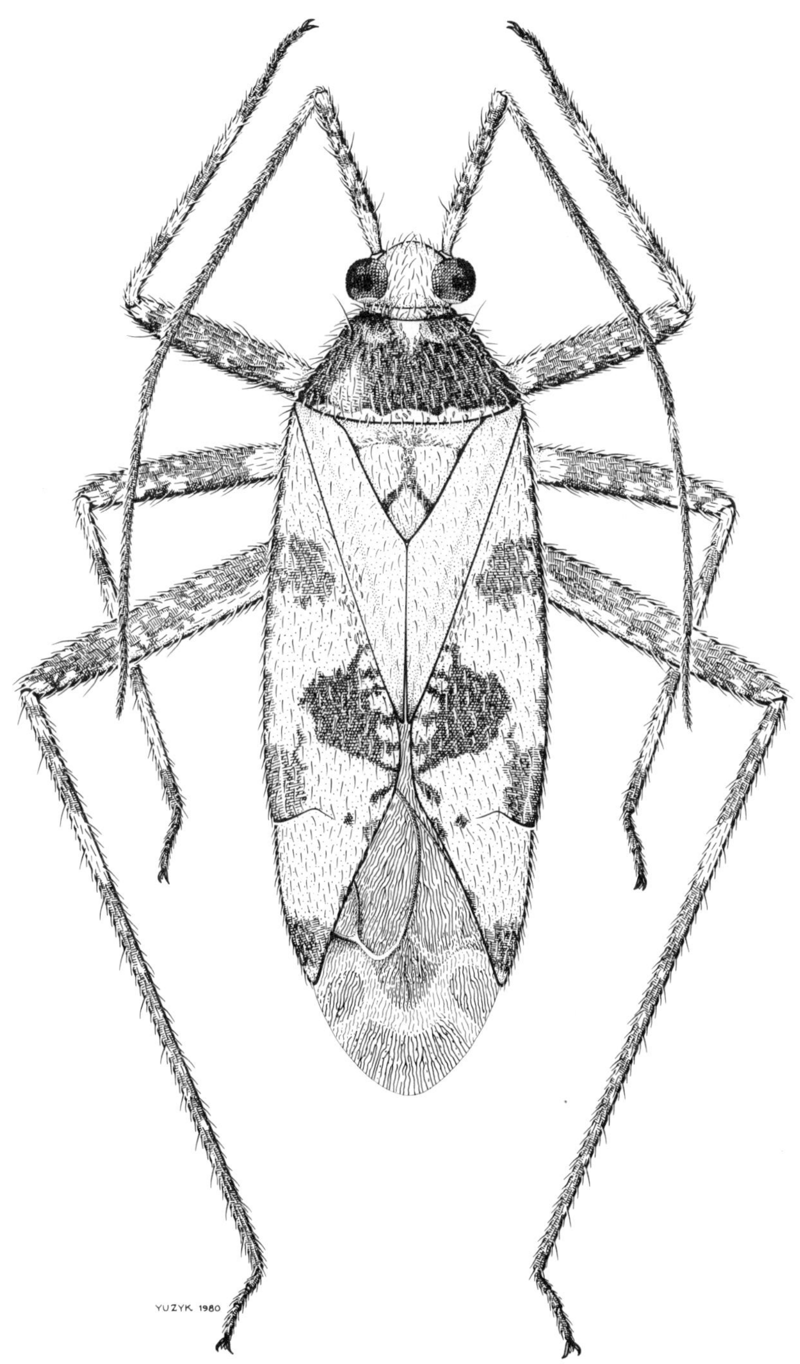

Fig. 56. *Phytocoris nigricollis*

Remarks.   This species is distinguished by the mottled appearance of the hemelytra and by the straight right clasper (Fig. 65). A short tubercle on the genital segment near the base of each clasper is present.

Collected on apple and pear in Quebec and Nova Scotia; predaceous on aphids and mites.

Distribution.   Minnesota, Ohio; Quebec (Map 29).

Phytocoris erectus Van Duzee

Fig. 66; Map 30

Phytocoris erectus Van Duzee, 1920:345.

Length 5.4–5.8 mm; width 1.9–2.2 mm. Head yellowish brown marked with reddish brown. Rostrum 2.5–2.7 mm long. First antennal segment pale on ventral surface, dorsal surface mottled with brown; second segment brown with pale band at base. Pronotum light brown, calli and collar marked with reddish brown. Hemelytra gray shaded with brown; apical area of corium with dark brown oblique area, and large gray area just behind; sericeous hairs silvery, simple hairs golden.

Remarks.   This species is distinguished by the dark brown oblique area on apical corium and by the right clasper (Fig. 66). A short tubercle on the genital segment above the left clasper is present.

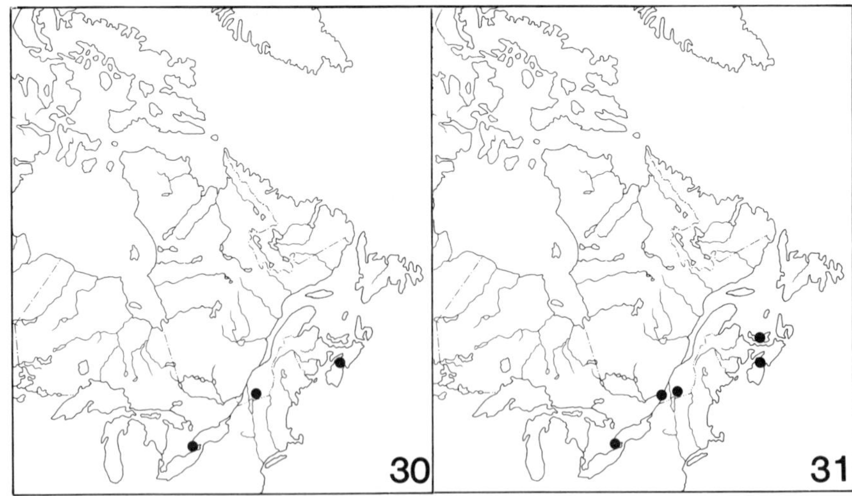

Map 30. Collection localities for *Phytocoris erectus.*

Map 31. Collection localities for *Phytocoris canadensis.*

80

Collected on apple in Nova Scotia, Quebec, and Ontario; predaceous on mites and aphids. MacLellan (1972) probably observed this species preying on codling moth larvae in Nova Scotia.

Also collected on *Salix* spp.

Distribution.   Eastern USA; Nova Scotia, Quebec, Ontario, Saskatchewan (Map 30).

Phytocoris canadensis Van Duzee

Fig. 67; Map 31

Phytocoris canadensis Van Duzee, 1920:346.

Length 5.2–5.7 mm; width 1.8–2.0 mm. Head yellowish brown marked with reddish brown. Rostrum 2.5–2.7 mm long. First antennal segment pale on ventral surface, dorsal surface mottled with reddish brown; second segment brown with pale band at base. Pronotum light brown, collar and calli marked with orange or red; subbasal margin with six black spots. Hemelytra gray shaded with brown; corium with markings and pubescence similar to that of *erectus*.

Remarks.   This species is similar to *erectus* in appearance but may be separated from it by differences in the claspers (Fig. 67). The right clasper has a long and slender basal process. There are no tubercles on the genital segment.

Collected on choke cherry in Prince Edward Island; on apple in Nova Scotia and Quebec; on apple, currant, and gooseberry in Ontario; predaceous on aphids. MacLellan (1972) probably observed this species preying on the codling moth eggs and larvae in Nova Scotia.

Distribution.   Eastern USA; Nova Scotia, Quebec, Ontario (Map 31).

Phytocoris salicis Knight

Figs. 57, 68; Map 32

Phytocoris salicis Knight, 1920:56.

Length 5.6–6.1 mm; width 2.1–2.3 mm. Head yellowish marked with brown. Rostrum 2.6–2.8 mm long. First antennal segment pale on ventral surface, dorsal surface mottled with brown; second segment mostly yellow, black near base and apex. Pronotum yellowish brown, collar and calli marked with red, lateral margins black, subbasal margin with several black spots. Scutellum mostly pale brown. Hemelytra light brown mottled with black; sericeous pubescence silvery, simple hairs golden.

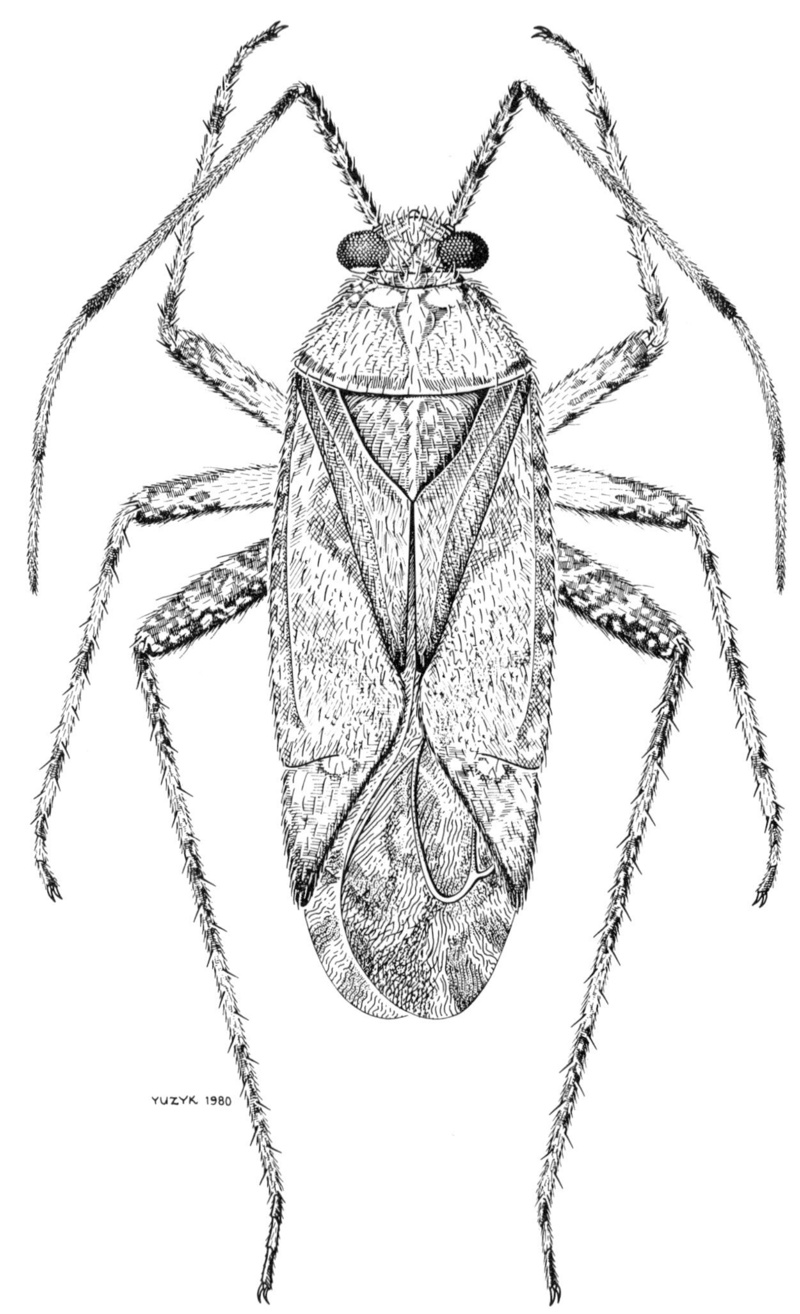

Fig. 57. *Phytocoris salicis*

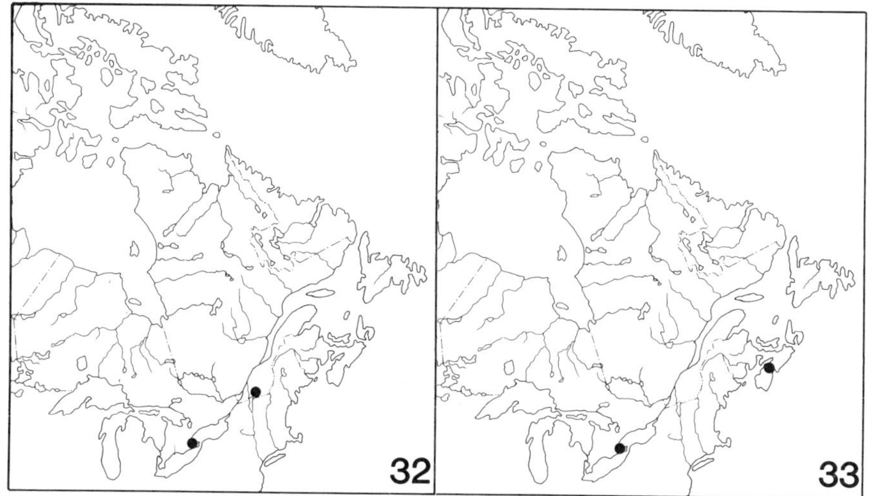

Map 32. Collection localities for *Phytocoris salicis.*
Map 33. Collection localities for *Phytocoris onustus.*

Remarks. This species is distinguished by the uniformly mottled hemelytra (Fig. 57) and by the differences in the claspers (Fig. 68). A short tubercle on the genital segment near the base of each clasper is present.

Collected on apple and pear in Quebec and Ontario; predaceous on aphids and mites.

Also collected on *Salix* spp.

Distribution. Eastern USA; Quebec, Ontario, Prairie Provinces (Map 32).

Phytocoris onustus Van Duzee

Fig. 69; Map 33

Phytocoris onustus Van Duzee, 1920:344.

Length 6.0–7.2 mm; width 2.3–2.5 mm. Head marked with black. Rostrum 2.9–3.1 mm long. First antennal segment pale on ventral surface, dorsal surface mottled with brown; second segment black with pale band at base. Pronotum brown, subbasal margin black interrupted by pale spaces. Hemelytra mottled with brown; inner apical area pale; sericeous hairs in clumps, silvery; simple hairs black.

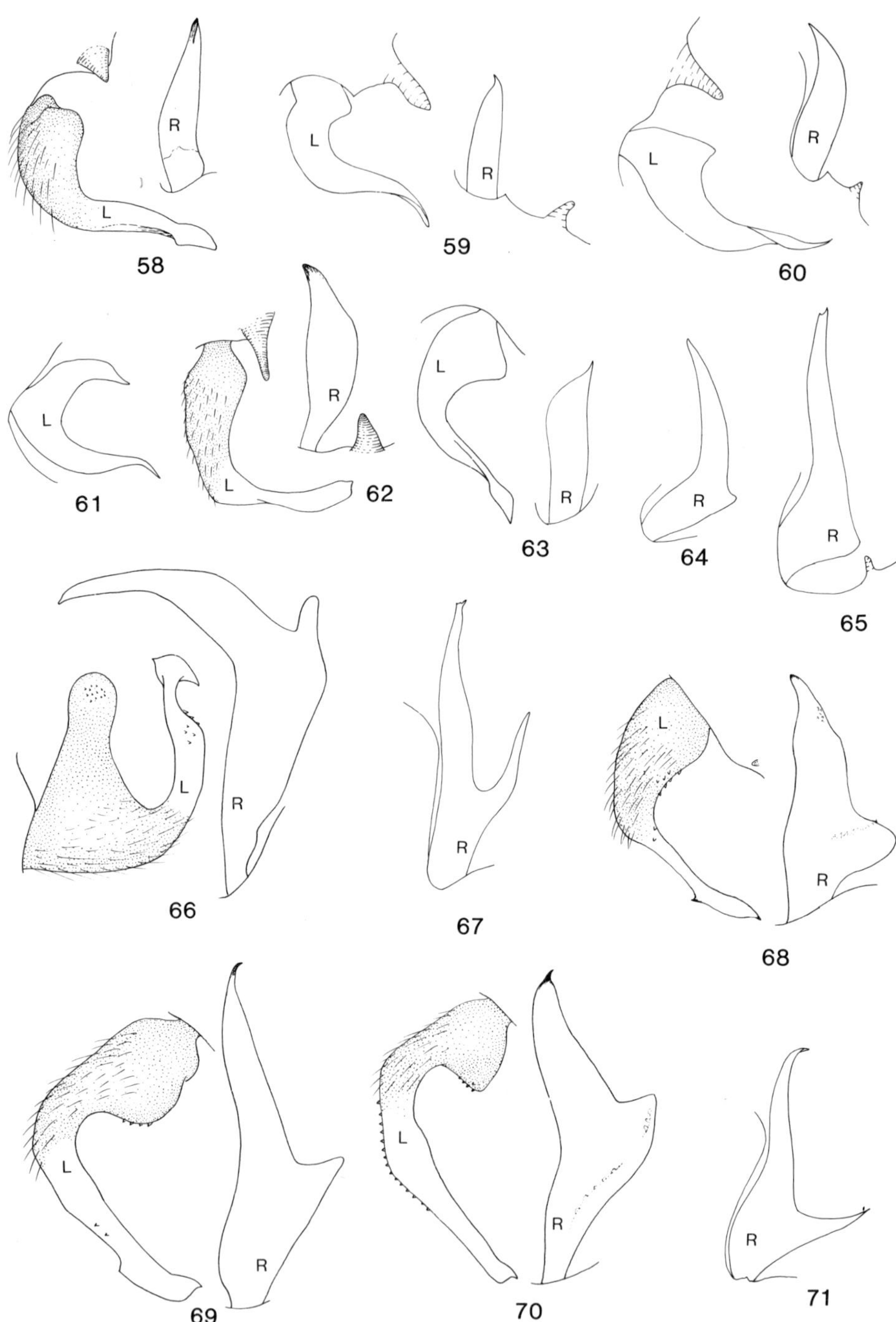

Figs. 58–71. Male claspers of *Phytocoris* spp. 58, *lasiomerus*; 59, *sulcatus*; 60, *cortice-vivens*; 61, *gracillatus*; 62, *conspurcatus*; 63, *dimidiatus*; 64, *nigricollis*; 65, *husseyi*; 66, *erectus*; 67, *canadensis*; 68, *salicis*; 69, *onustus*; 70, *neglectus*; 71, *cortitectus*.

Remarks.   This species is distinguished by the long rostrum, by the pale band at the base of second antennal segment, and by the claspers (Fig. 69). There are no tubercles on the genital segment.

Collected on apple and pear in Nova Scotia; on apple, pear, and mulberry in Ontario, usually on trunks hiding in crevices; predaceous on small arthropods.

Distribution.   Eastern USA; Nova Scotia, Ontario, Manitoba (Map 33).

Phytocoris neglectus Knight

Fig. 70; Map 34

Phytocoris neglectus Knight, 1920:30.

Length 6.1–6.5 mm; width 2.1–2.3 mm. Head brown marked with darker brown. Rostrum 2.8–2.9 mm long. First antennal segment pale on ventral surface, dorsal surface mottled with brown; second segment brown with pale band at base. Pronotum light brown, subbasal margin darker brown. Scutellum light brown, diagonal bar near middle brown. Hemelytra light brown mottled with brown, oblique bar on apical corium brown; sericeous hairs silvery, simple hairs black.

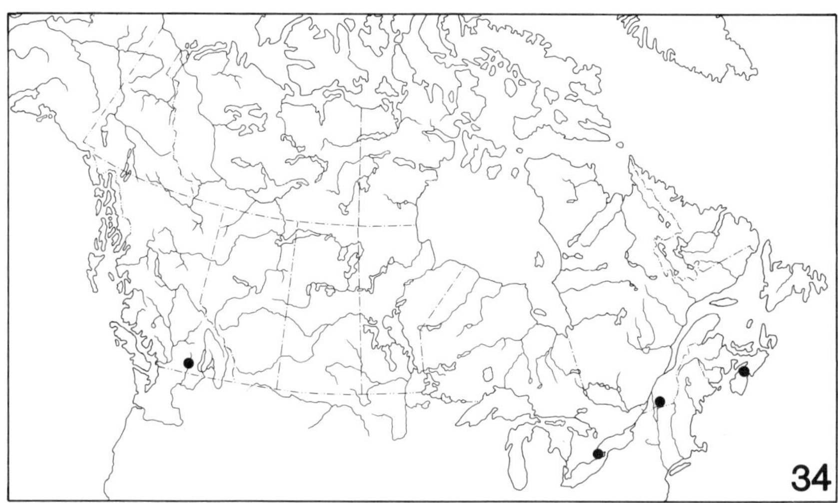

Map 34. Collection localities for *Phytocoris neglectus.*

Remarks. This species is distinguished by the oblique brown mark on the apical corium and by the claspers (Fig. 70). There are no tubercles on the genital segment.

Collected on apple and pear in Nova Scotia, Quebec, and Ontario; on apple, pear, peach, and sweet cherry in British Columbia; predaceous on aphids, psyllids, and mites. Knight (1941*b*) observed the species preying on psocids in New York.

Also collected on *Picea glauca* and *Abies balsamea*.

Distribution. Eastern USA; Nova Scotia, Quebec, Ontario, Prairie Provinces, British Columbia (Map 34).

Phytocoris cortitectus Knight

Fig. 71; Map 35

Phytocoris cortitectus Knight, 1920:55.

Length 6.0–6.2 mm; width 2.0–2.2 mm. Head marked with brown. Rostrum 2.8–3.0 mm long. First antennal segment pale on ventral surface, dorsal surface mottled with brown; second segment brown with pale band at base. Pronotum light brown, subbasal margin with four connected black spots; collar marked with reddish orange. Scutellum yellow, oblique bar each side near apex reddish brown. Hemelytra mottled with brown; apex of corium with large opaque spot.

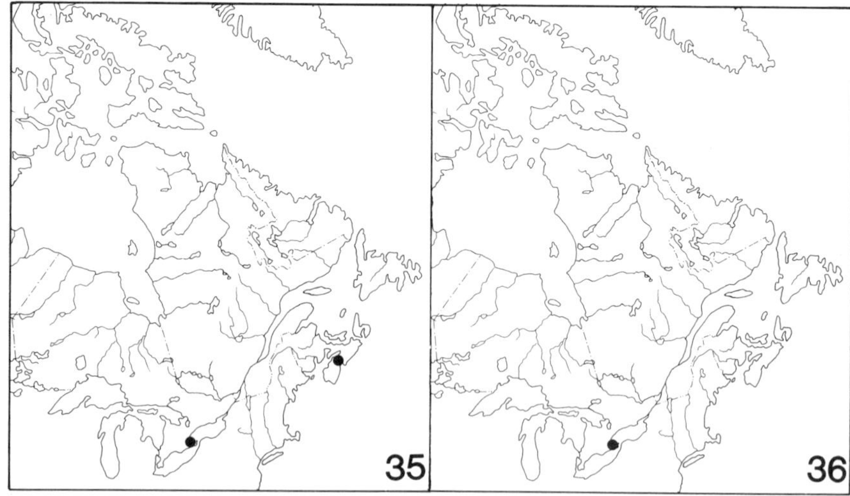

Map 35. Collection localities for *Phytocoris cortitectus*.
Map 36. Collection locality for *Ceratocapsus pilosulus*.

Remarks. This species is distinguished by the long rostrum and by the right clasper (Fig. 71). There are no tubercles on the genital segment.

Collected on apple and pear in Nova Scotia; predaceous on aphids and mites.

Distribution. Eastern USA; Nova Scotia, Ontario (Map 35).

Subfamily Orthotylinae Van Duzee

The following are the subfamily characteristics: 1) large, free parempodia, converging at apices; 2) small and depressed collar; and 3) flexible ductus seminis, without membranous lobes, with or without spicula.

The subfamily is represented by 2 tribes, 9 genera, and 16 species. Fourteen species are predaceous, two are phytophagous.

Key to tribes of Orthotylinae

1. Hemelytra without transverse bands of silvery, sericeous pubescence . . . . . .
   . . . . **Orthotylini** (p. 87)
   Hemelytra with transverse bands of silvery, sericeous pubescence (Fig. 88) . . .
   . . . . **Pilophorini** (p. 116)

Tribe Orthotylini

The tribe is represented by eight genera and 13 species. Eleven species are predaceous, two are phytophagous.

Key to genera of Orthotylini

1. Antennal segments 3 and 4 thickened, as thick as apex of second (Figs. 72–77)
   . . . . *Ceratocapsus* **Reuter** (p. 88)
   Antennal segments 3 and 4 thinner, thinner than apex of second segment . .
   2
2. Second antennal segment greatly inflated (Fig. 78) . . . .
   . . . . *Heterotoma* **Le Peletier & Serville** (p. 99)
   Second antennal segment not inflated 3
3. Hemelytra with scaly pubescence (Figs. 79,88) . . . .
   . . . . *Heterocordylus* **Fieber** (p. 101)
   Hemelytra without scaly pubescence 4
4. Red species with black legs (Fig. 81) . . . . *Lopidea* **Uhler** (p. 104)
   Green species with green legs 5
5. Head black (Fig. 82) . . . . *Paraproba* **Distant** (p. 104)
   Head green or yellow 6

6. Pronotum with basal angles black; bases of tibiae with black spots (Fig. 83)
.... *Blepharidopterus* **Kolenati** (p. 106)
Pronotum all green; bases of tibiae without black spots 7
7. Delicate slender species; eyes near middle of head (Fig. 84) ....
.... *Diaphnocoris* **Kelton** (p. 110)
Robust species; eyes adjacent to pronotum (Figs. 85,86) ....
.... *Orthotylus* **Fieber** (p. 112)

Genus *Ceratocapsus* Reuter

Elongate or oval, dark brown species. Head oblique, basal margin overlaps apical portion of pronotum; eyes prominent, carina between them distinct. Antennae stout, nearly of equal thickness throughout. Pronotum and hemelytra smooth or punctate; pubescence simple or simple intermixed with sericeous hairs.

The genus contains many species that are similar in appearance. The males may be identified by the differences in the claspers and the females by association with the males.

Six species were collected. Overwinter in the egg stage. The nymphs appear in early June and the adults in early July. The adults are active throughout July and August, and gradually die out by the end of August. The species are predaceous, feeding on many small arthropods found on the plants.

Key to species *Ceratocapsus*

1. Hemelytra with wide, pale transverse band (Fig. 72); claspers (Fig. 89) ....
.... *pilosulus* **Knight** (p. 89)
Hemelytra without pale transverse band 2
2. Pronotum and hemelytra impunctate, with short and long simple hairs (Fig. 73); claspers (Fig. 90) .... *modestus* **(Uhler)** (p. 89)
Pronotum and hemelytra punctate, with simple and sericeous hairs 3
3. Hind femora pale yellow 4
Hind femora brown or red 5
4. Clavus mostly brown; simple hairs on hemelytra dense (Fig. 74); claspers (Fig. 91) .... *digitulus* **Knight** (p. 92)
Clavus mostly pale yellow; simple hairs on hemelytra not as dense (Fig. 75); right clasper (Fig. 92) .... *incisus* **Knight** (p. 94)
5. Species dark brown, cuneus brown (Fig. 76); right clasper (Fig. 93) ....
.... *pumilus* **(Uhler)** (p. 96)
Species reddish brown, cuneus reddish (Fig. 77); right clasper (Fig. 94) ...
.... *fuscinus* **Knight** (p. 97)

Ceratocapsus pilosulus Knight

Figs. 72, 89; Map 36

Ceratocapsus pilosus Knight, 1923*b*:526 (n. preoc.).
Ceratocapsus pilosulus Knight, 1930*a*:198.

Length 3.2–3.5 mm; width 1.3–1.6 mm. Head, pronotum, and scutellum brown. Hemelytra brown with wide, pale transverse band.

Remarks. This species is distinguished by the pale transverse band on the hemelytra (Fig. 72) and by the claspers (Fig. 89).

Collected on apple, pear, apricot, plum, peach, and mulberry in Ontario; predaceous on aphids.

Also collected on *Ostrya virginiana*, *Corylus americana*, and *Quercus macrocarpa*.

Distribution. Northeastern and central USA; Manitoba, Ontario (Map 36).

Ceratocapsus modestus (Uhler)

Figs. 73, 90; Map 37

Melinna modesta Uhler, 1887:69.
Ceratocapsus modestus: Smith, 1909:161.

Length 4.3–4.5 mm; width 1.6–1.8 mm. Head light to dark brown. Antennae brown. Pronotum and scutellum dark brown. Hemelytra light to dark brown, impunctate; pubescence simple, with short and long hairs. Legs brown.

Remarks. This species is distinguished by the brown color, by the impunctate pronotum and hemelytra, by the short and long, simple pubescence (Fig. 73), and by the claspers (Fig. 90).

Collected on apple and pear in Quebec; on apple, pear, plum, grape, and mulberry in Ontario; predaceous on white flies.

Also collected on *Alnus rugosa*, *Quercus macrocarpa*, *Tilia americana*, *Juglans nigra*, *Ulmus americana*, and *Salix* spp.

Distribution. Eastern and central USA; Saskatchewan, Manitoba, Ontario, Quebec (Map 37).

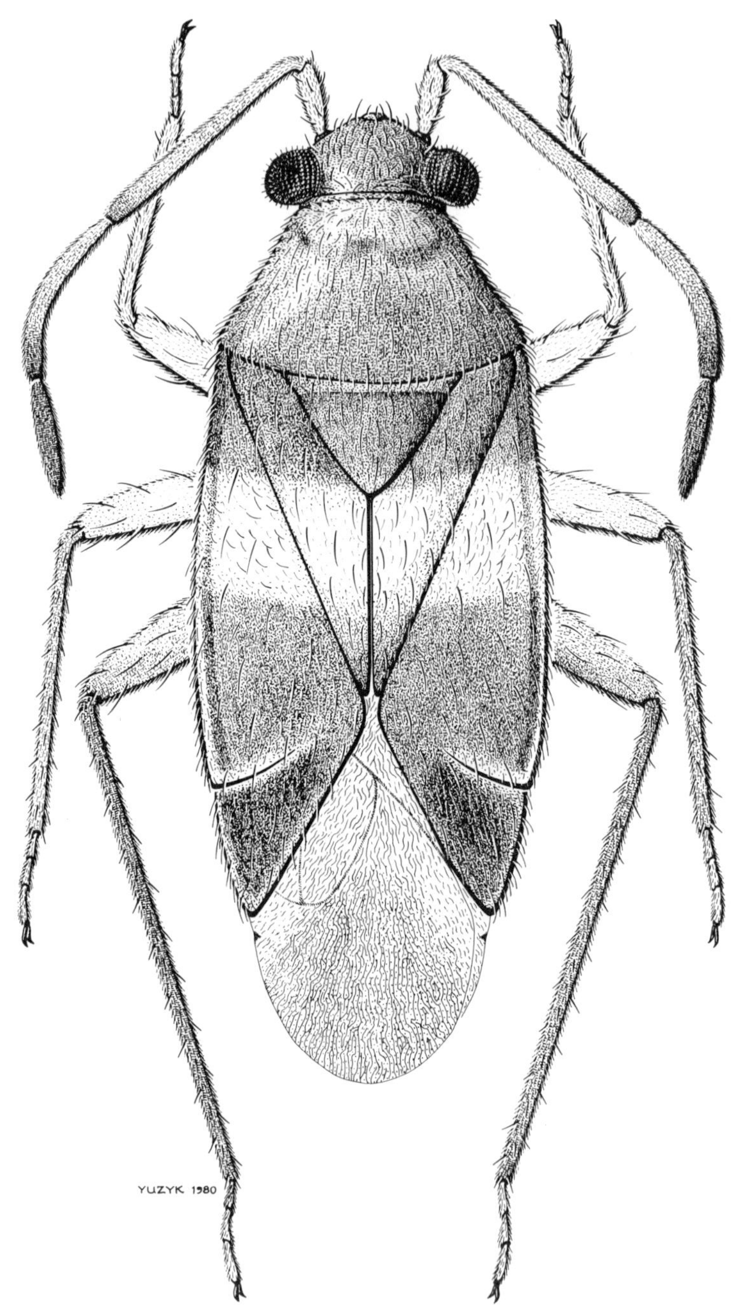

Fig. 72. *Ceratocapsus pilosulus*

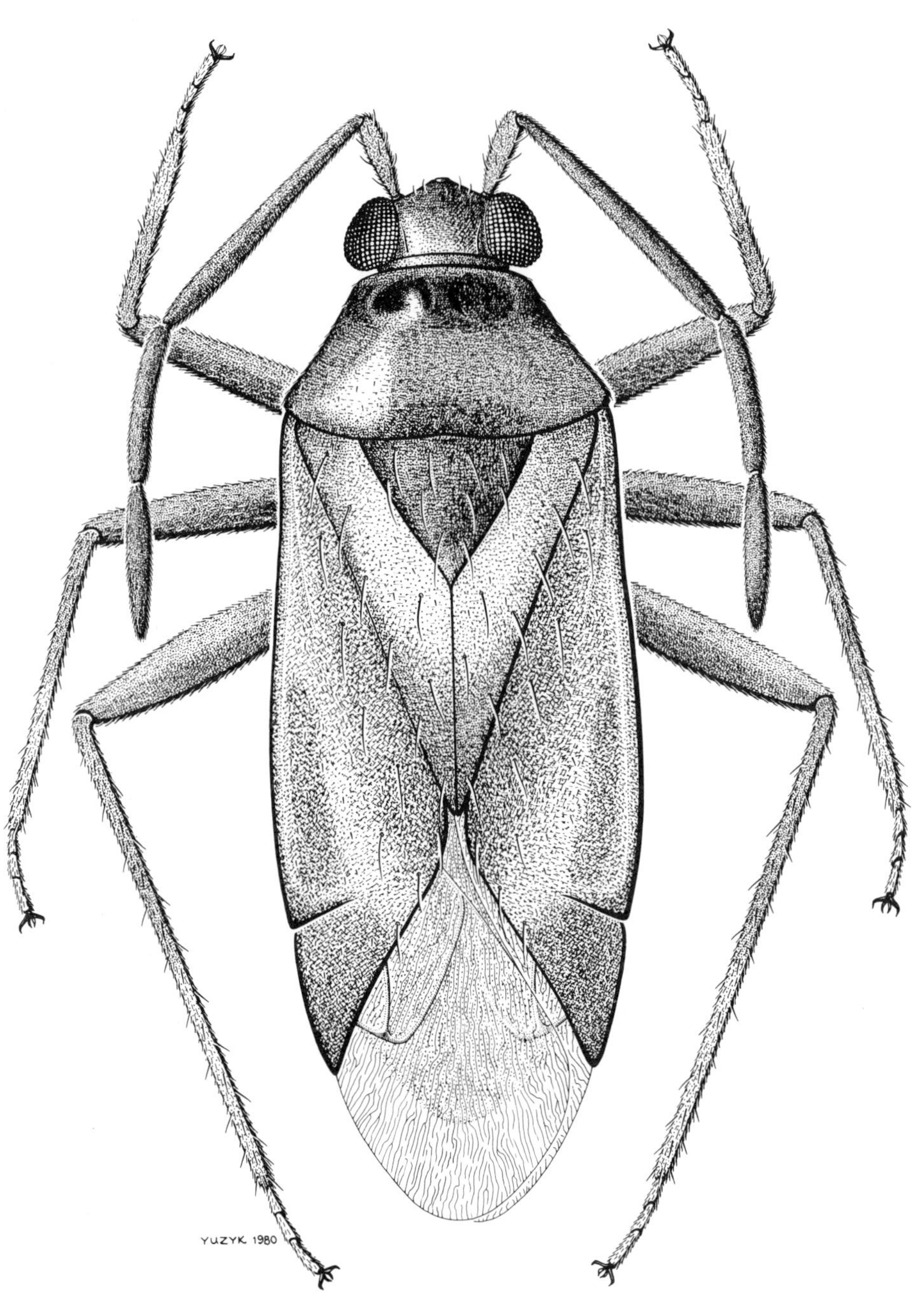

Fig. 73. *Ceratocapsus modestus*

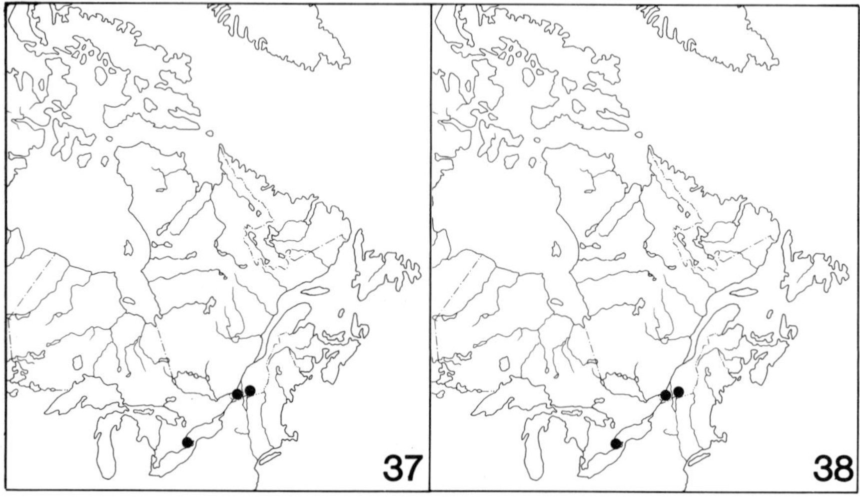

Map 37. Collection localities for *Ceratocapsus modestus*.
Map 38. Collection localities for *Ceratocapsus digitulus*.

Ceratocapsus digitulus Knight

Figs. 74, 91; Map 38

Ceratocapsus digitulus Knight, 1923*b*:522.

Length 3.5–3.8 mm; width 1.6–1.8 mm. Head brown; first and second antennal segments pale yellow, first segment with red bar near base. Pronotum and hemelytra brown, punctate; simple pubescence dense, intermixed with appressed sericeous hairs. Legs pale yellow.

Remarks. This species is distinguished by the pale first and second antennal segments, by the dense, simple pubescence on the hemelytra (Fig. 74), and by the claspers (Fig. 91).

Collected on apple and pear in Quebec; on apple, pear, peach, and plum in Ontario; predaceous on aphids and mites.

Also collected on *Salix interior*.

Distribution. Eastern and central USA; Manitoba, Ontario, Quebec (Map 38).

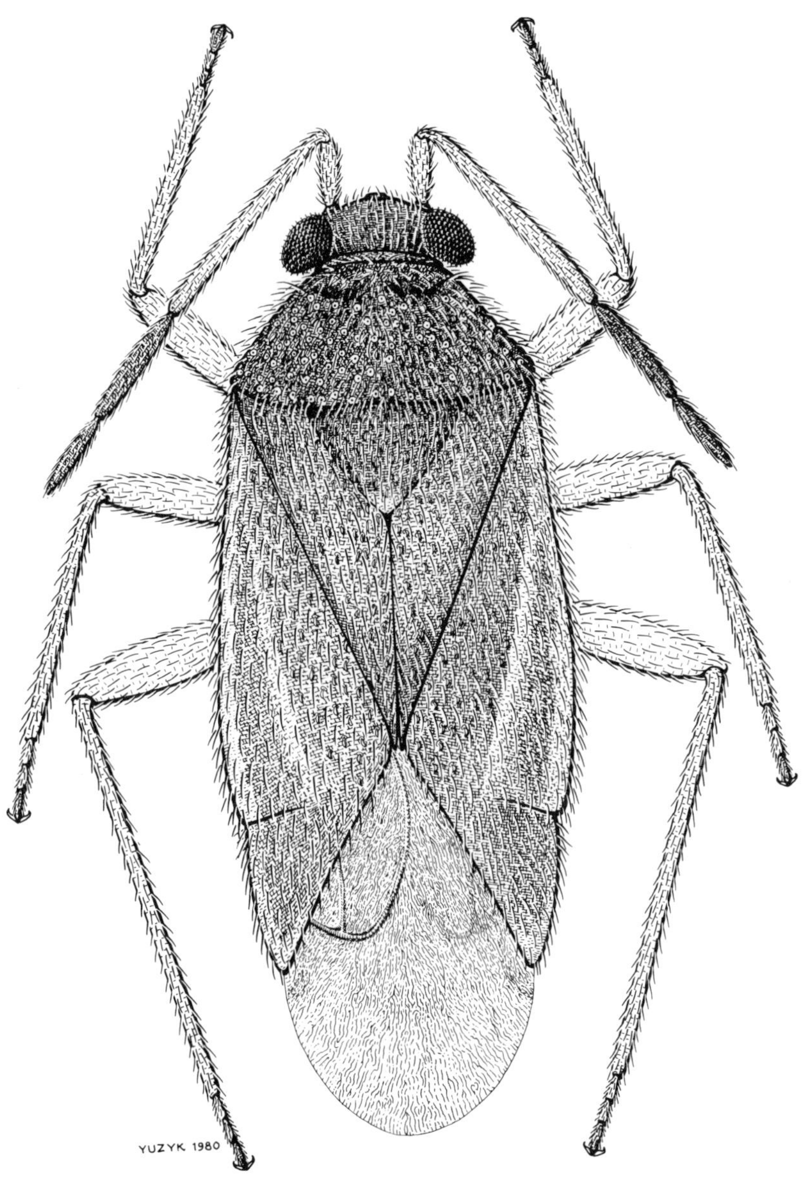

Fig. 74. *Ceratocapsus digitulus*

Ceratocapsus incisus Knight

Figs. 75, 92; Map 39

Ceratocapsus incisus Knight, 1923*b*:532.

Length 3.7–4.0 mm; width 1.51–7.0 mm. Head reddish brown. First and second antennal segments pale yellow, first segment with red bar near base. Pronotum and scutellum brown. Hemelytra brown; clavus and costal margin paler; pubescence similar to that of *digitulus*. Legs pale yellow.

Remarks.   This species is distinguished by the pale yellow first and second antennal segments, by the pale yellow legs (Fig. 75), and by the right clasper (Fig. 92). The simple pubescence on the hemelytra is not as dense as in *digitulus*.

Collected on apple, pear, peach, apricot, and wild and cultivated grape in Ontario; predaceous on aphids.

Also collected on *Alnus rugosa*, *Carpinus caroliniana*, and *Salix* spp.

Distribution.   Eastern USA; Ontario (Map 39).

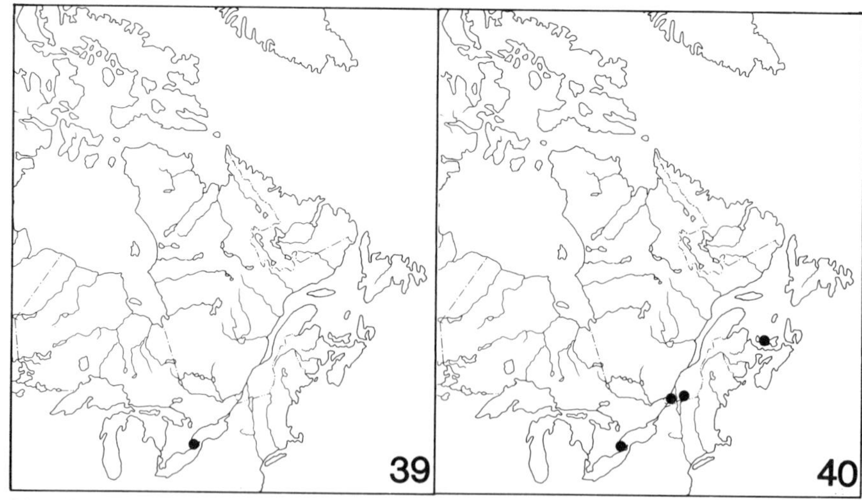

Map 39. Collection locality for *Ceratocapsus incisus*.
Map 40. Collection localities for *Ceratocapsus pumilus*.

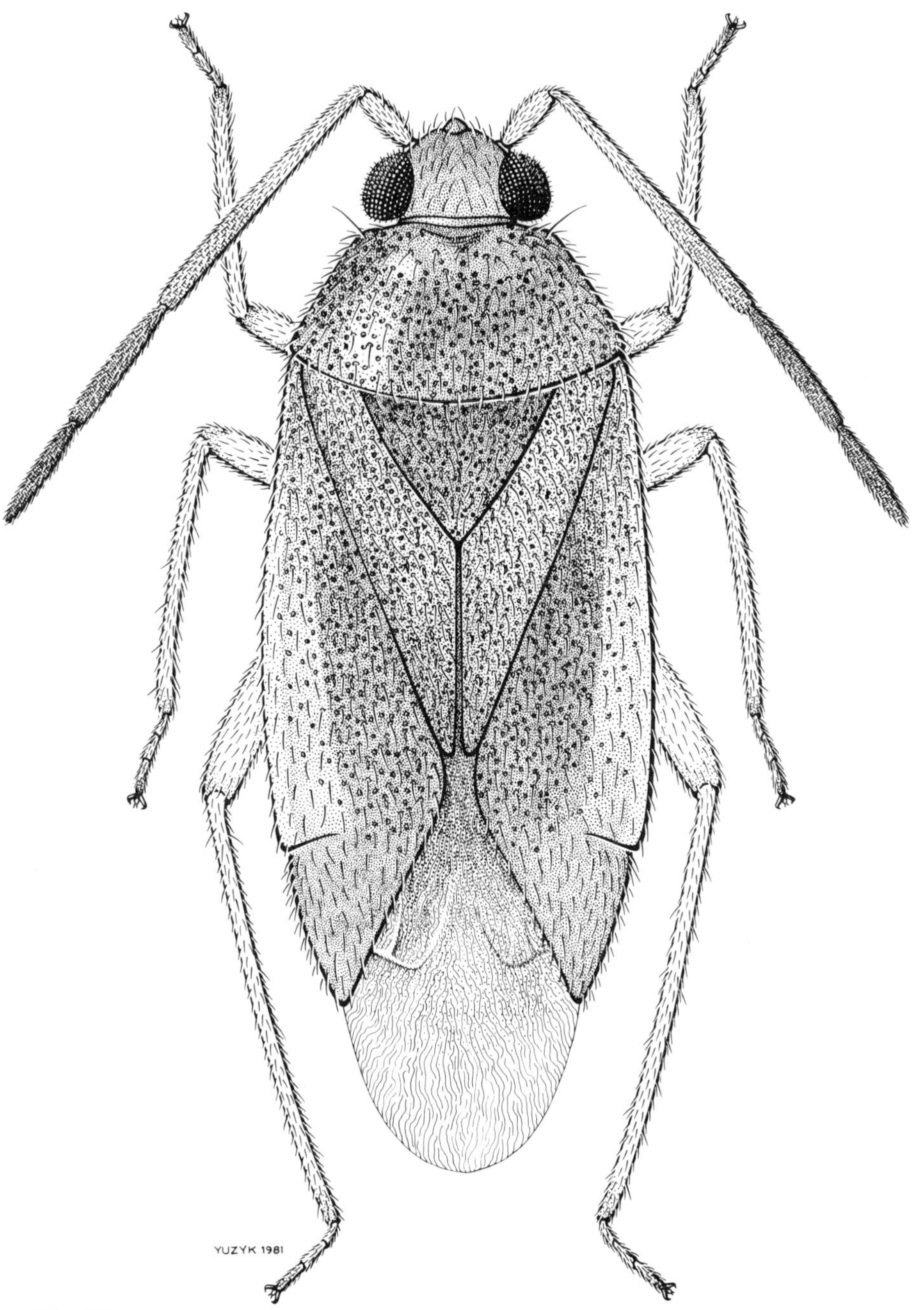

Fig. 75. *Ceratocapsus incisus*

Ceratocapsus pumilus (Uhler)

Figs. 76, 93; Map 40

Melinna pumila Uhler, 1887:69.
Ceratocapsus pumilus: Van Duzee, 1909:182.

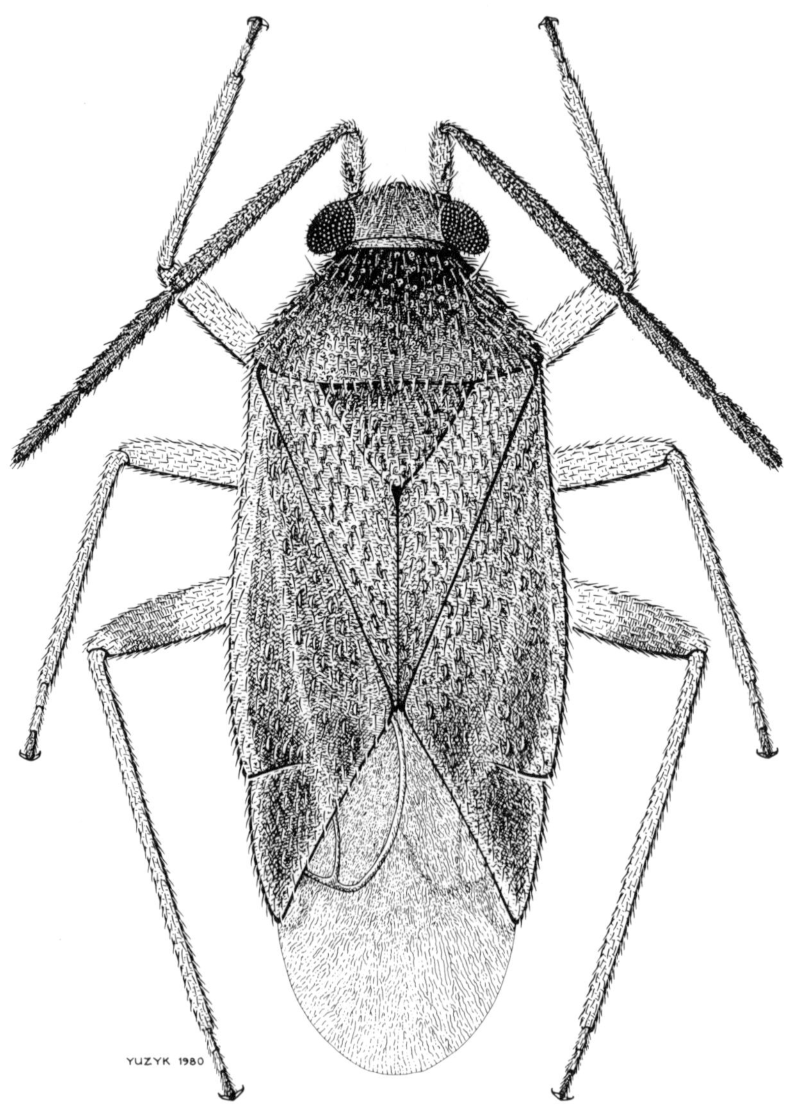

Fig. 76. *Ceratocapsus pumilus*

Length 3.7–3.9 mm; width 1.5–1.7 mm. Head brown; first and second antennal segments light brown, first segment with red bar near base. Pronotum, scutellum, and hemelytra brown; pubescence similar to that of *incisus*. Legs light brown, apical half of hind femur often dark brown.

Remarks. This species is distinguished by the brown color (Fig. 76) and by the right clasper (Fig. 93).

Collected on raspberry in Prince Edward Island; on apple and pear in Quebec; on apple, pear, peach, and wild grape in Ontario; predaceous on aphids and mites.

Also collected on *Betula nigra* and *Salix* spp.

Distribution. Eastern USA; Prince Edward Island, Quebec, Ontario (Map 40).

Ceratocapsus fuscinus Knight

Figs. 77, 94; Map 41

Ceratocapsus fuscinus Knight, 1923*b*:531.

Length 3.6–3.7 mm; width 1.5–1.6 mm. Head yellow; first antennal segment yellow with red bar near base, second segment yellow. Pronotum light brown, basal margin yellow. Hemelytra light brown, cuneus tinged with red; pubescence similar to that of *pumilus*. Legs yellow, apical half of hind femur reddish.

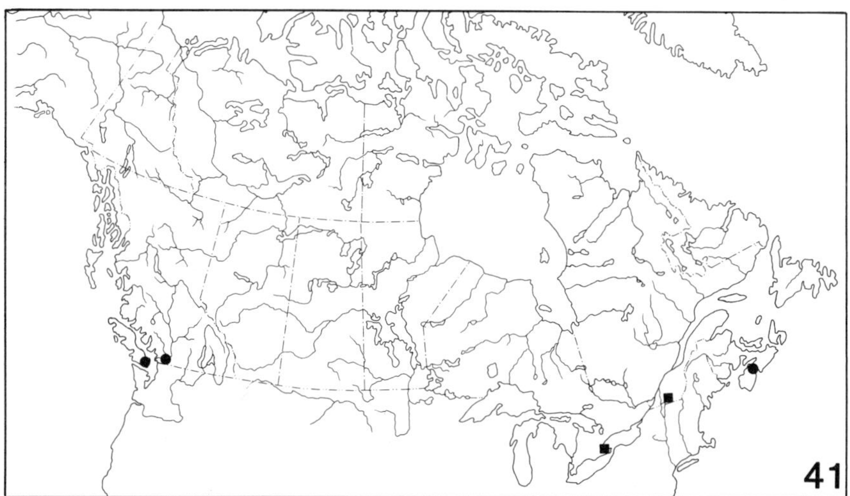

Map 41. Collection localities for *Ceratocapsus fuscinus* (■), and *Heterotoma meriopterum* (●).

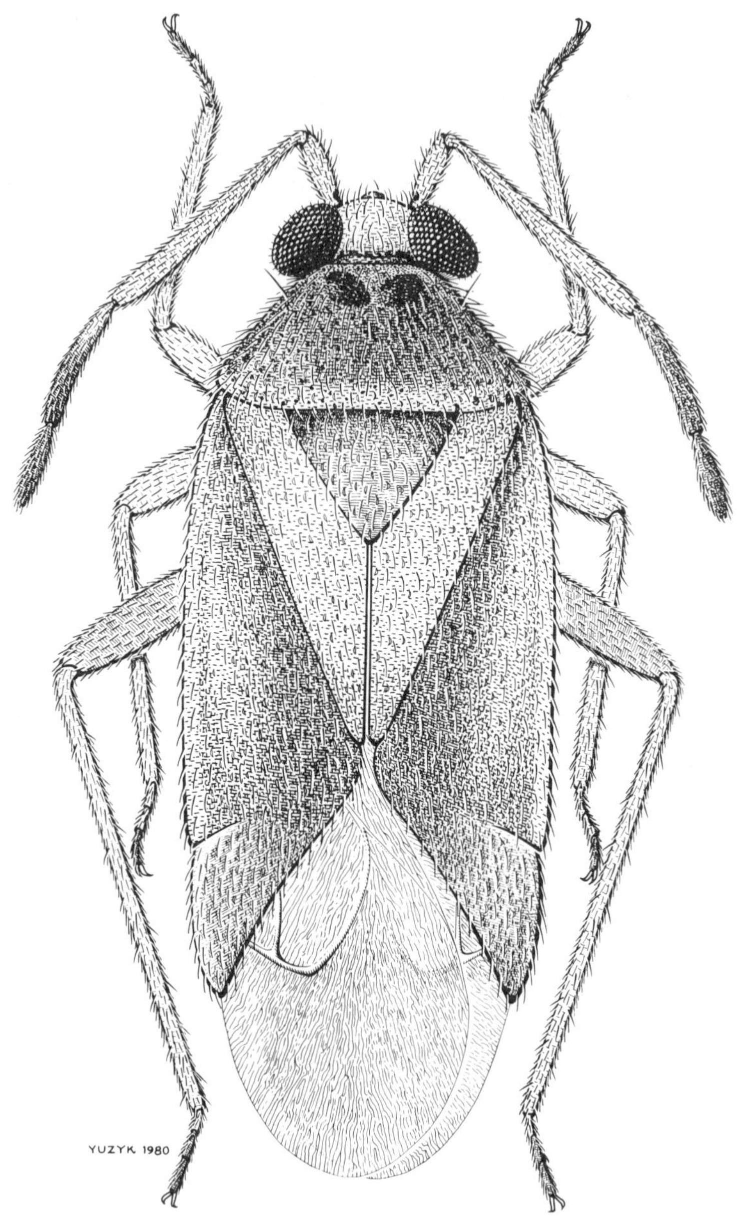

Fig. 77. *Ceratocapsus fuscinus*

Remarks. This species is similar to *pumilus* in appearance but is more yellowish, with reddish hind femur and reddish cuneus (Fig. 77). The right clasper (Fig. 94) is also different.

Collected on apple, pear, and wild grape in Quebec; on apple, pear, peach, apricot, and plum in Ontario; predaceous on aphids and mites.

Also collected on *Salix* spp.

Distribution. Eastern USA; Ontario, Quebec (Map 41).

Genus *Heterotoma* Le Peletier & Serville

Elongate, black species. Head oblique; eyes spherical, carina between them distinct. Second antennal segment greatly inflated. Pronotum and hemelytra faintly rugose; pubescence simple and sericeous.

One species, introduced from Europe, was collected. Overwinters in the egg stage.

Heterotoma meriopterum (Scopoli)

Fig. 78; Map 41

Cimex meriopterum Scopoli, 1763:131.
Heterotoma meriopterum: Fieber, 1861:290.

Length 5.0–5.2 mm; width 1.4–1.5 mm. Antennae black, second segment greatly inflated. Ventral surface black, legs pale green.

Remarks. Knight (1917*a*) first reported this European species from New York, and Downes (1957) from British Columbia. It is distinguished by the black color and by the greatly inflated second antennal segment (Fig. 78).

Collected on raspberry in Nova Scotia; on apple and raspberry in British Columbia; predaceous on aphids.

Also collected on *Corylus maxima* and *Cytisus scoparius.*

The nymphs appear in June and the adults in July. The adults are active throughout August and gradually die out in September.

Distribution. New York; Nova Scotia, British Columbia (Map 41).

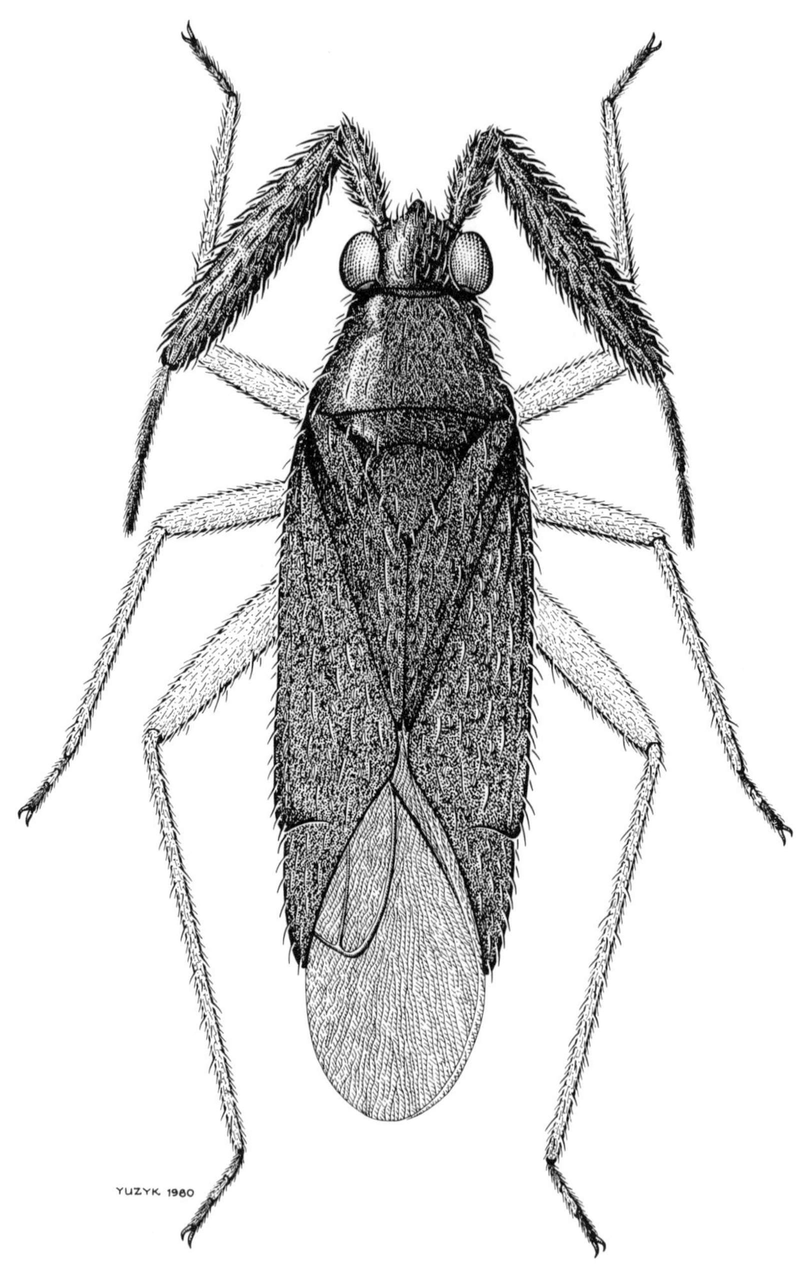

Fig. 78. *Heterotoma meriopterum*

Genus *Heterocordylus* Fieber

Robust, black or black and red species. Head oblique, base of head sharply truncate; eyes large, carina between them distinct. Pronotum rugose, side margins carinate. Scutellum rugose. Hemelytra impunctate; pubescence simple and scalelike.

One species was collected. Overwinters in the egg stage.

Heterocordylus malinus Reuter

Figs. 79, 80; Map 42

Heterocordylus malinus Reuter, 1909:71.

Length 6.3–7.0 mm; width 2.1–2.3 mm. Head black, spot on vertex next to eye often red; antennae black. Pronotum black in males, mostly red in females. Scutellum black. Hemelytra and abdomen mostly black in males, black and red in females. Legs black.

Remarks. This species is distinguished by the large size, by the black (Fig. 79) or black and red color (Fig. 80), and by the white scalelike pubescence.

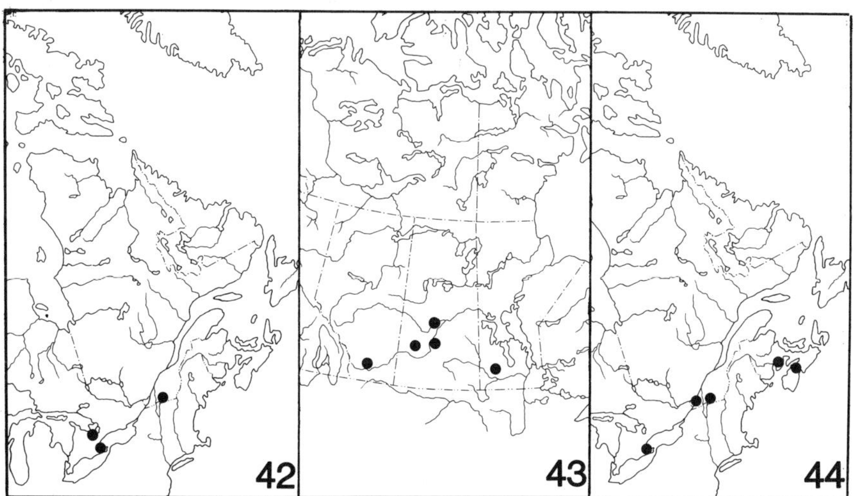

Map 42. Collection localities for *Heterocordylus malinus*.
Map 43. Collection localities for *Lopidea dakota*.
Map 44. Collection localities for *Paraproba capitata*.

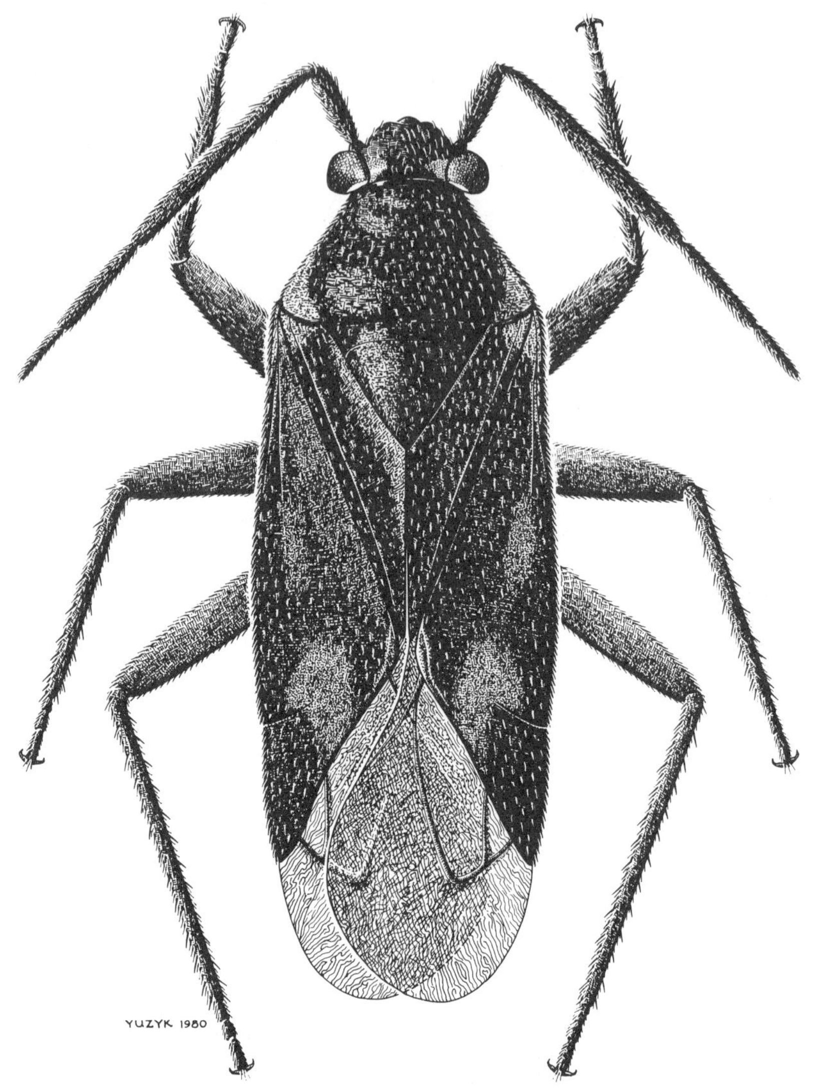

Fig. 79. *Heterocordylus malinus,* ♂

Collected on apple in Ontario and Quebec; phytophagous. Caesar (1912) and Knight (1918) reported the species as a pest of apple in Ontario and New York, respectively.

The nymphs appear about the first of May and the adults in early June. By mid-July most of the adults die out.

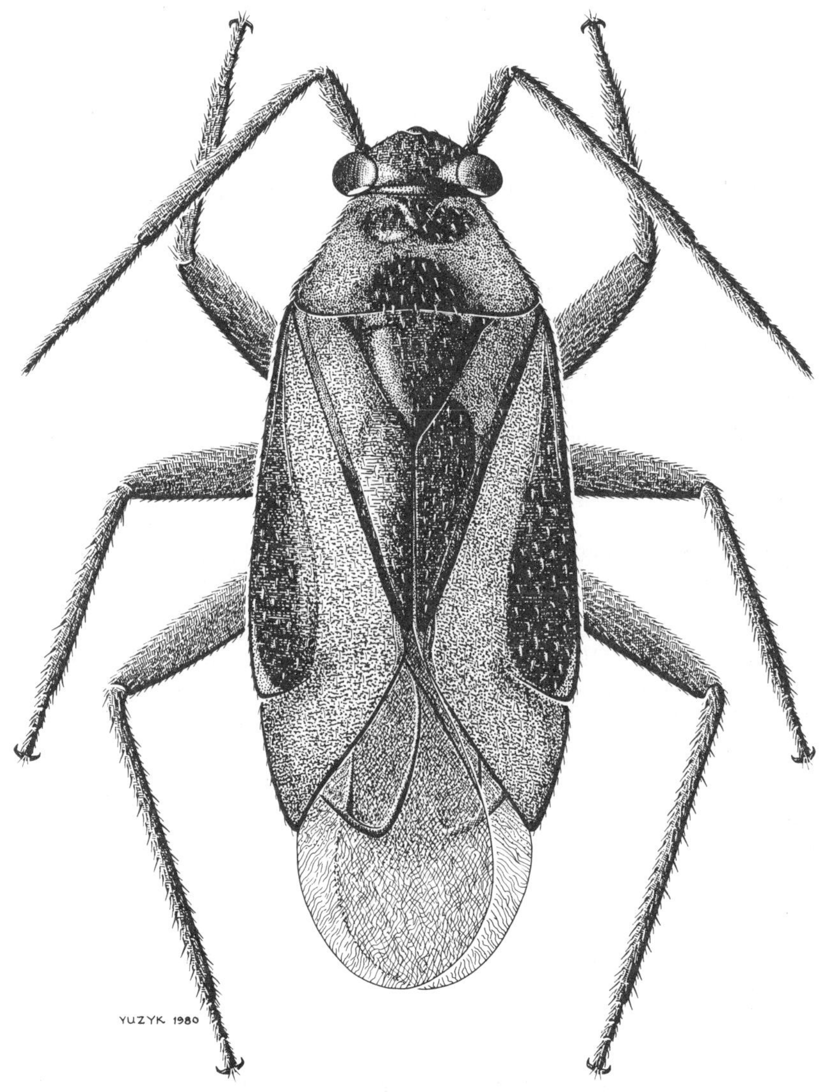

Fig. 80. *Heterocordylus malinus,* ♀

Also breeds on *Crataegus chrysocarpa*; adults readily migrate to apple trees and feed on the fruit, especially if the apple trees are nearby.

Distribution.   Eastern USA; Quebec, Ontario (Map 42).

Genus *Lopidea* Uhler

Elongate, red species. Head vertical; eyes spherical, carina between them distinct. Antennae black. Hemelytra impunctate; pubescence simple, black. Legs black.

One species was collected. Overwinters in the egg stage.

Lopidea dakota Knight

Figs. 81, 95; Map 43

Lopidea dakota Knight, 1923*a*:67.

Length 6.3–6.5 mm; width 2.3–2.5 mm. Head red, frons often black. Pronotum red. Scutellum black. Hemelytra red; narrow inner margin of clavus and wing membrane black.

Remarks. This species is distinguished by the large size, by the red and black color (Fig. 81), and by the right clasper (Fig. 95).

Collected on raspberry in Manitoba; phytophagous. King and Glen (1936) and Twinn (1939) reported the species damaging raspberry in Saskatchewan and Alberta, respectively; Glen and King (1938) and Arneson et al. (1939) reported it damaging currant and strawberry, respectively, in Saskatchewan.

The nymphs appear in early June and the adults in early July. The adults are active throughout July and August, and gradually die out by early September.

Also breeds on *Caragana arborescens*.

Distribution. North central USA; British Columbia, Prairie Provinces (Map 43).

Genus *Paraproba* Distant

Slender, pale green delicate species. Head ventrical; eyes large, carina between them absent. Pronotum and hemelytra impunctate; pubescence simple.

One species was collected. Overwinters in the egg stage.

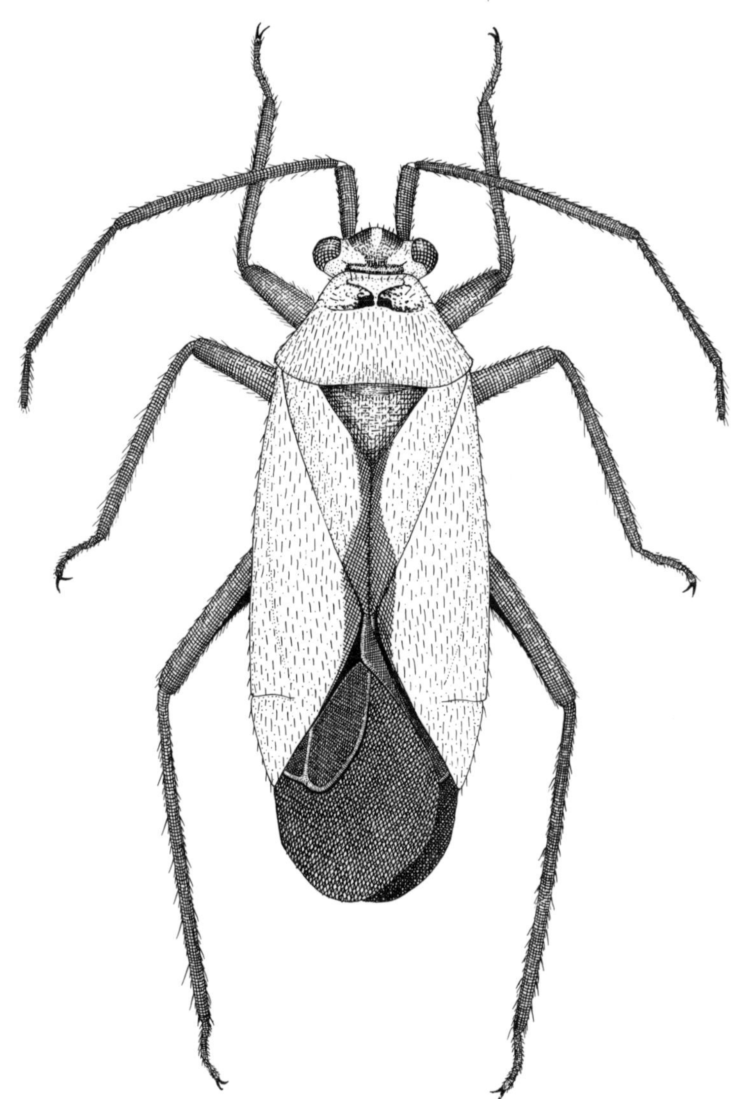

Fig. 81. *Lopidea dakota*

Paraproba capitata (Van Duzee)

Fig. 82; Map 44

Diaphnidia capitata Van Duzee, 1912:490.
Diaphnocoris capitata: Kelton, 1961:566.
Paraproba capitata: Kelton, 1965:1028.

Length 3.0–3.5 mm; width 0.9–1.1 mm. Head black; first antennal segment black. Pronotum and hemelytra light green. Legs green.

Remarks.   This species is distinguished by the black head (Fig. 82).

Collected on apple and pear in Nova Scotia and Quebec; on apple in New Brunswick; on apple, pear, and mulberry in Ontario; predaceous on mites, aphids, and leafhoppers. Gilliatt (1935) reported the species preying on the red mite in Nova Scotia.

The nymphs appear about the end of May and the adults at the end of June. The adults are active throughout July and August, and gradually die out by early September.

Also collected on *Zanthoxylum americanum, Ulmus americana, Corylus americana, Carya* spp., *Fraxinus* spp., *Crataegus* spp., *Hamamelis virginiana,* and *Robinia pseudoacacia.*

Distribution.   Northeastern and north central USA; Nova Scotia, New Brunswick, Quebec, Ontario (Map 44).

Genus *Blepharidopterus* Kolenati

Elongate, green species. Head oblique; eyes large, carina between them distinct. Pronotum and hemelytra impunctate; pubescence simple, black.

One species, introduced from Europe, was collected. Overwinters in the egg stage.

Blepharidopterus angulatus Fallén

Fig. 83; Map 45

Lygaeus angulatus Fallén, 1807:76.
Blepharidopterus angulatus: Kirkaldy, 1906:128.

Length 5.1–5.3 mm; width 1.4–1.5 mm. Head yellowish green. First antennal segment yellowish green, often with a longitudinal black line.

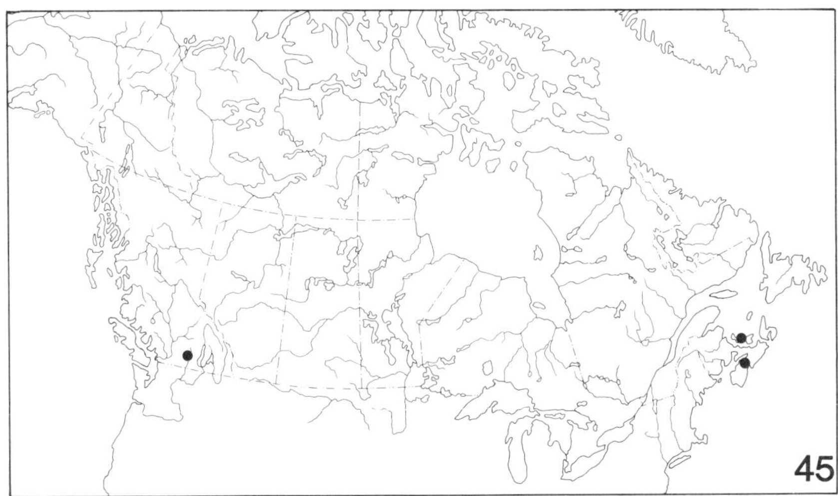

Map 45. Collection localities for *Blepharidopterus angulatus*.

Pronotum green, basal angles black. Hemelytra green. Legs yellowish green, bases of tibiae with black spots.

Remarks. Knight (1921*b*) first reported this European species from Nova Scotia, and Downes (1927) from British Columbia. It is distinguished by the black basal angles on the pronotum and by the black spot at the base of each tibia (Fig. 92).

Collected on apple and plum in Prince Edward Island; on apple, pear, and plum in Nova Scotia; on apple, pear, peach, sweet cherry, and sour cherry in British Columbia; predaceous on aphids, psyllids, mites, and other small arthropods. Lord (1971) reported the species preying on the eggs of the red mite, and codling moth eggs and young larvae in Nova Scotia.

The nymphs appear about mid-May and the adults about mid-June. The adults are active throughout July and August, and gradually die out by early September.

Also collected on *Alnus rugosa*, *Betula papyrifora*, *Fagus sylvatica*, *Fraxinus excelsior*, *Acer saccharum*, *Tilia cordata*, and *Ulmus americana*.

Distribution. Prince Edward Island, Nova Scotia, British Columbia (Map 45).

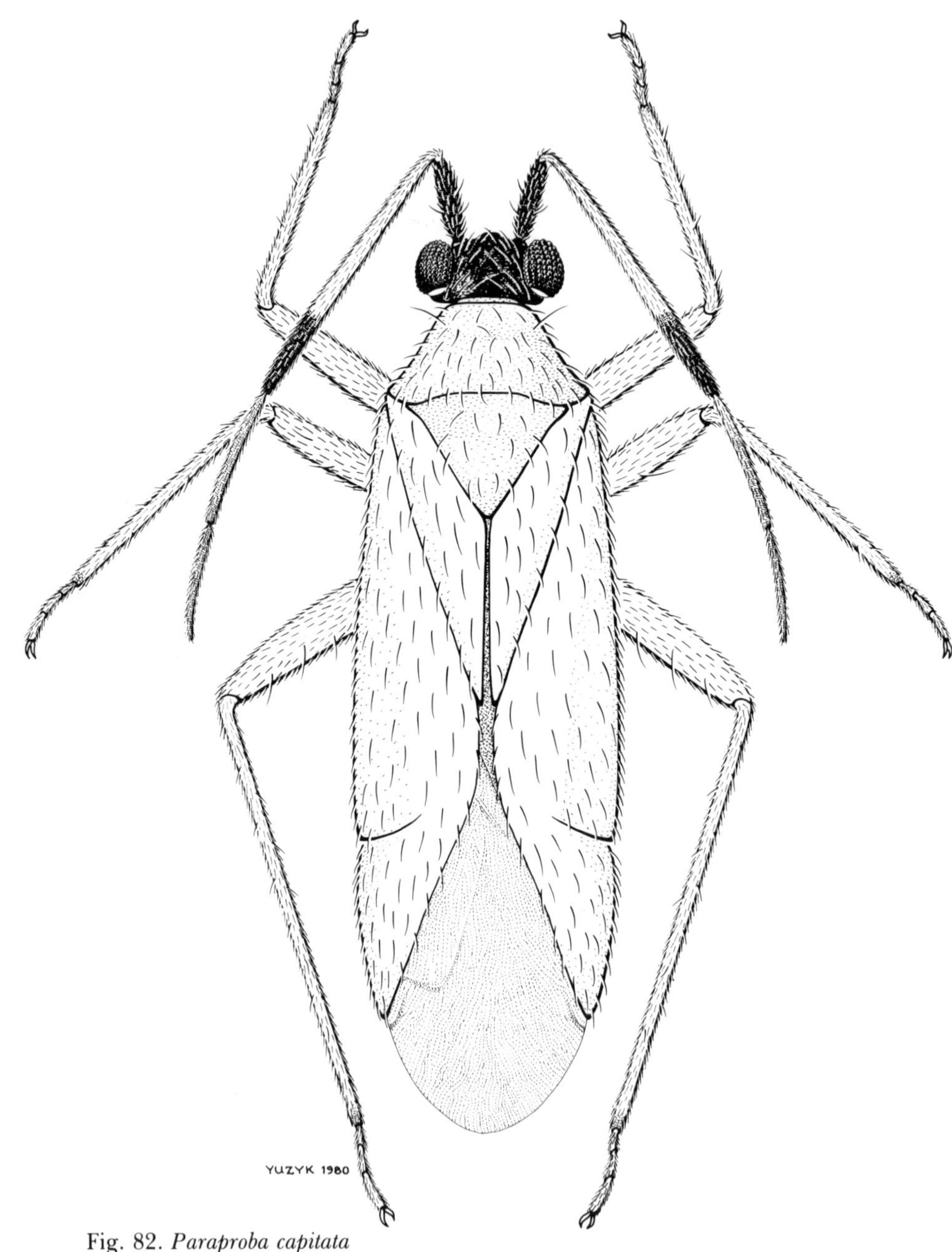

Fig. 82. *Paraproba capitata*

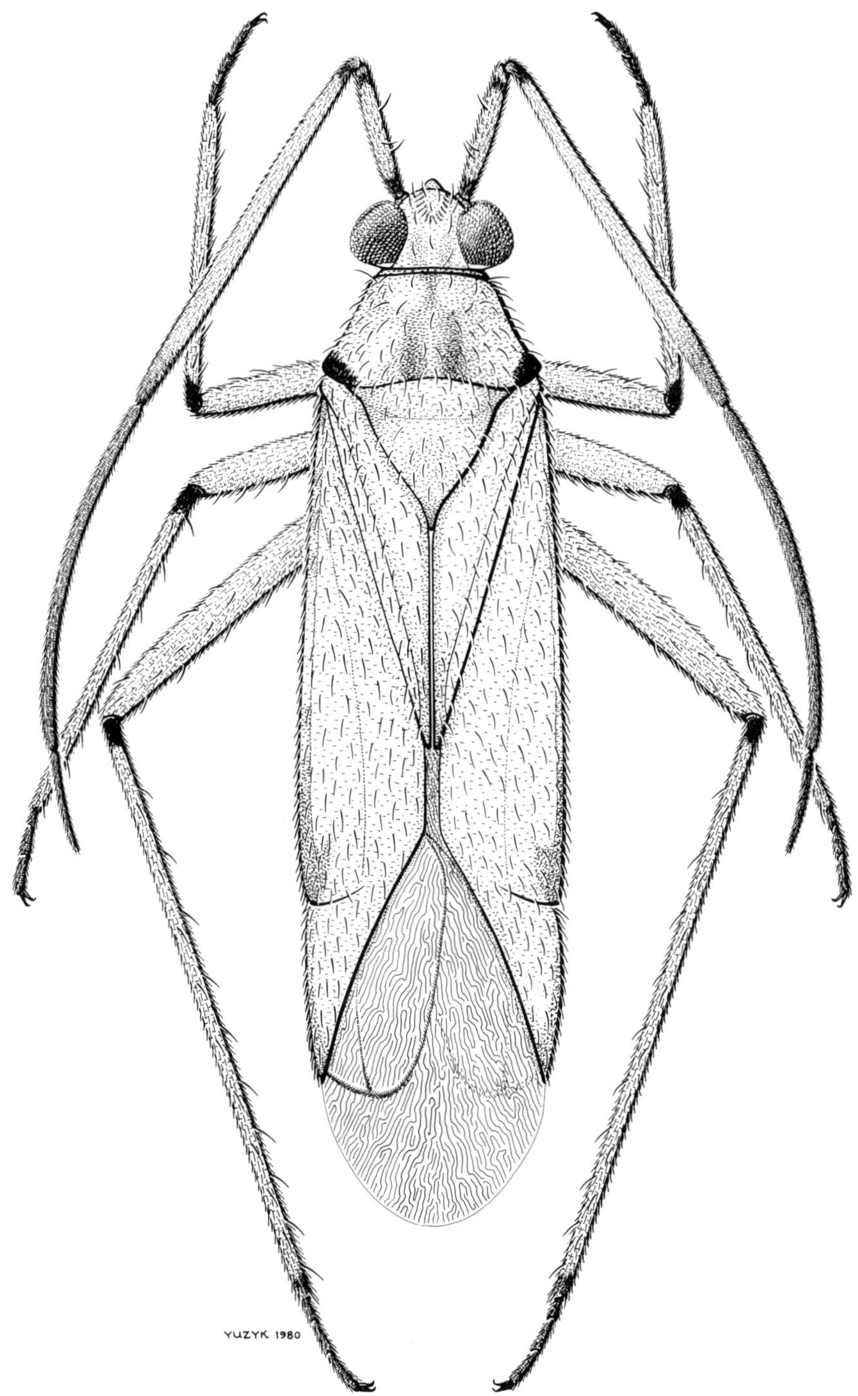

Fig. 83. *Blepharidopterus angulatus*

Genus *Diaphnocoris* Kelton

Slender, pale green species. Head oblique; eyes large, carina between them indistinct. Pronotum and hemelytra impunctate; pubescence simple. Legs green.

One species was collected. Overwinters in the egg stage.

Diaphnocoris provancheri (Burque)

Fig. 84; Map 46

Melacocoris provancheri Burque, *in* Provancher, 1887:144.
Diaphnocoris provancheri: Kelton, 1980*b*:343.

Length 4.2–4.7 mm; width 1.4–1.5 mm. Head pale green; eyes situated forward from posterior margin of head. Second antennal segment green or fuscous. Pronotum and hemelytra pale green. Legs pale green, hind tibia often fuscous.

Remarks. This species is distinguished by the position of the eyes on the head (Fig. 84).

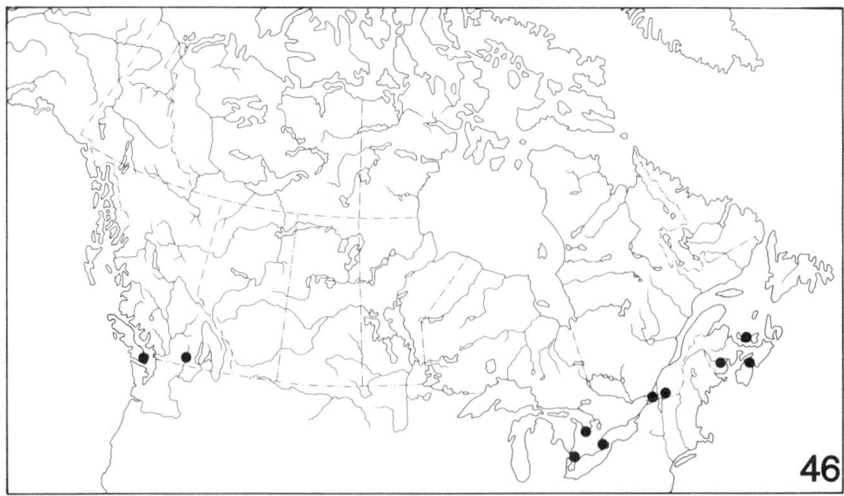

Map 46. Collection localities for *Diaphnocoris provancheri.*

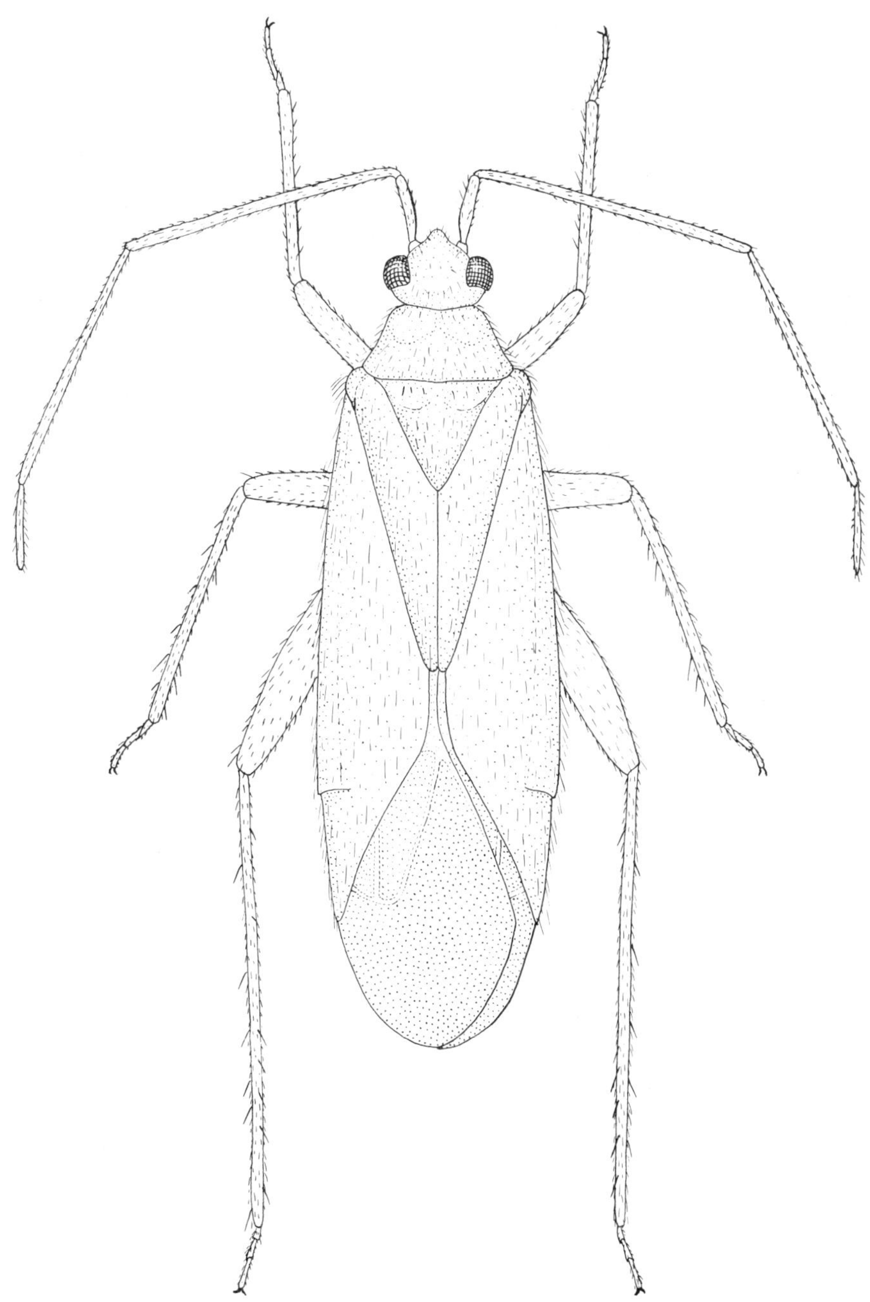

Fig. 84. *Diaphnocoris provancheri*

Collected on apple in Prince Edward Island, New Brunswick, and Quebec; on apple, pear, and plum in Nova Scotia; on apple, pear, peach, sour cherry, sweet cherry, plum, apricot, and mulberry in Ontario; on apple, pear, peach, and sweet cherry in British Columbia; predaceous on mites, aphids, and other small arthropods. Gilliatt (1930) reported the species preying on the red mite, MacPhee and Sanford (1954) reported the species preying on mites and aphids, MacLellan (1972) reported the species preying on codling moth eggs and young larvae in Nova Scotia, and McMullen and Jong (1967) reported the species preying on mites and pear psylla in British Columbia.

The nymphs appear about mid-May and the adults about mid-June. The adults are active throughout July and August, and gradually die out by the middle of September.

Also collected on *Acer* spp., *Alnus rugosa*, *Betula papyrifora*, *Corylus americana*, *Crataegus* spp., *Quercus* spp., and *Shepherdia canadensis*.

Distribution.   Widespread in USA; transcontinental in Canada (Map 46).

Genus *Orthotylus* Fieber

Elongate, green species. Head oblique, carina between eyes distinct. Pronotum smooth. Hemelytra impunctate; pubescence simple.

Two species, introduced from Europe, were collected. Overwinter in the egg stage. The nymphs appear about mid-May and the adults about mid-June. The adults are active throughout July and August, and gradually die out by early September.

Key to species of *Orthotylus*

1. First antennal segment uniformly pale green, longer than width of vertex on head; segment with many long bristles (Fig. 86); right claspers (Fig. 96) .... *viridinervis* (**Kirschbaum**) (p. 113)
   First antennal segment with black on ventral surface, shorter than width of vertex on head, segment with only few long bristles (Fig. 87); left clasper (Fig. 97) .... *nassatus* (**Fabricius**) (p. 115)

Orthotylus viridinervis (Kirschbaum)

Figs. 85, 96; Map 47

Capsus viridinervis Kirschbaum, 1855:238.
Orthotylus viridinervis: Fieber, 1861:290.

Length 5.6–5.8 mm; width 1.7–1.9 mm. Head green; first and second antennal segments green, terminal segments light brown. Rostrum 1.6–1.7 mm long, extending to tips of middle coxae. Pronotum and hemelytra green. Veins on wing membrane green. Legs green.

Remarks. Henry and Wheeler (1979) first reported this European species from Ontario, and Kelton (1982) from Nova Scotia. It is distinguished by the uniformly green color, by the long first antennal segment with many long bristles (Fig. 85), and by the right clasper (Fig. 96).

Collected on apple and pear in Nova Scotia; on apple in Ontario; predaceous on aphids.

Also collected on *Ulmus americana, Fagus sylvatica, Fraxinus excelsior,* and *Tilia cordata.*

Distribution. Europe; Nova Scotia, Ontario (Map 47).

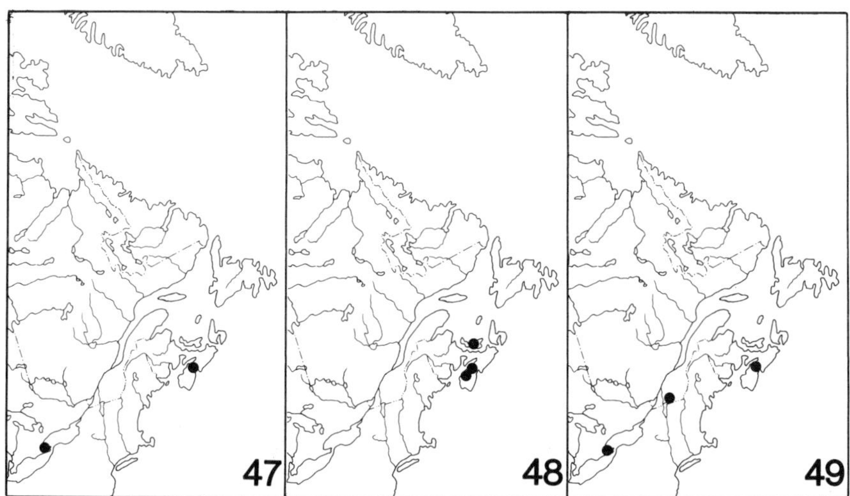

Map 47. Collection localities for *Orthotylus viridinervis.*
Map 48. Collection localities for *Orthotylus nassatus.*
Map 49. Collection localities for *Pilophorus perplexus.*

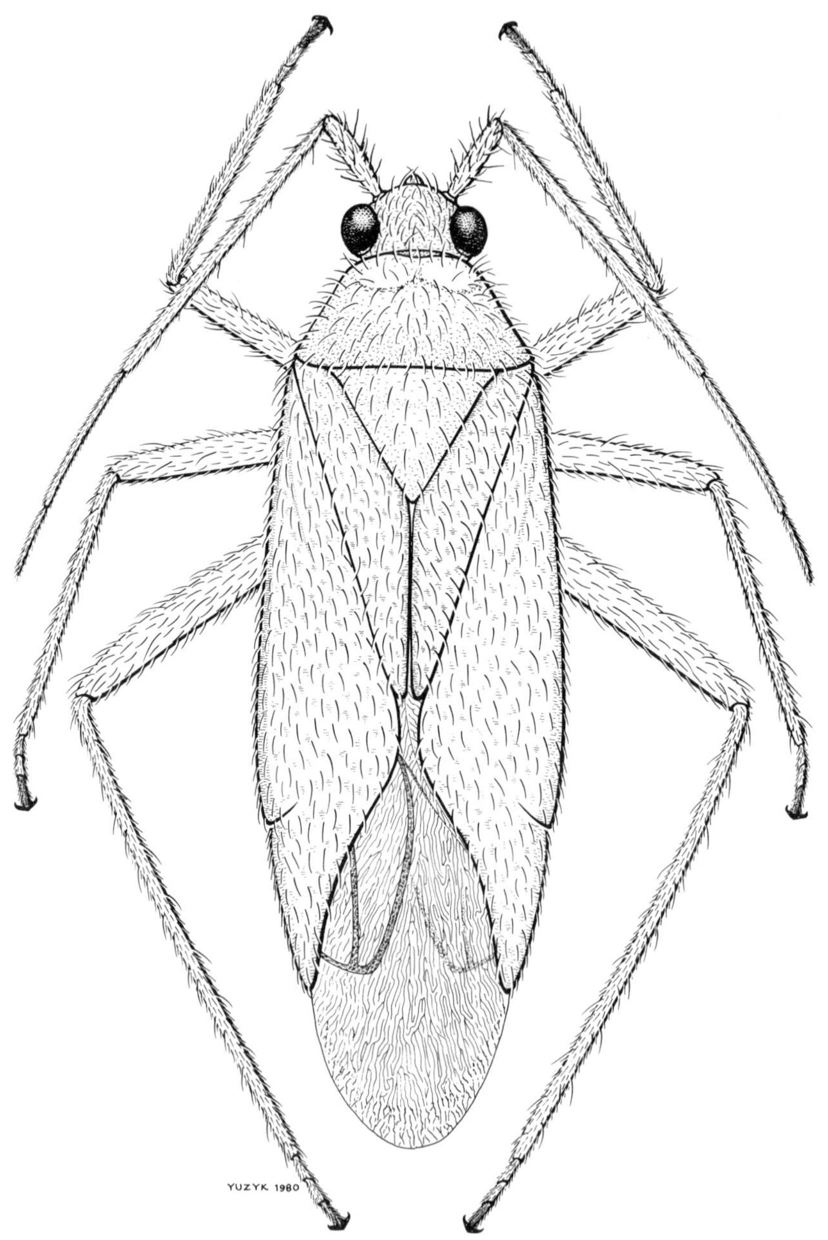

Fig. 85. *Orthotylus viridinervis*

114

Orthotylus nassatus (Fabricius)

Figs. 86, 97; Map 48

Cimex nassatus Fabricius, 1787:304.
Orthotylus nassatus: Fieber, 1861:289.

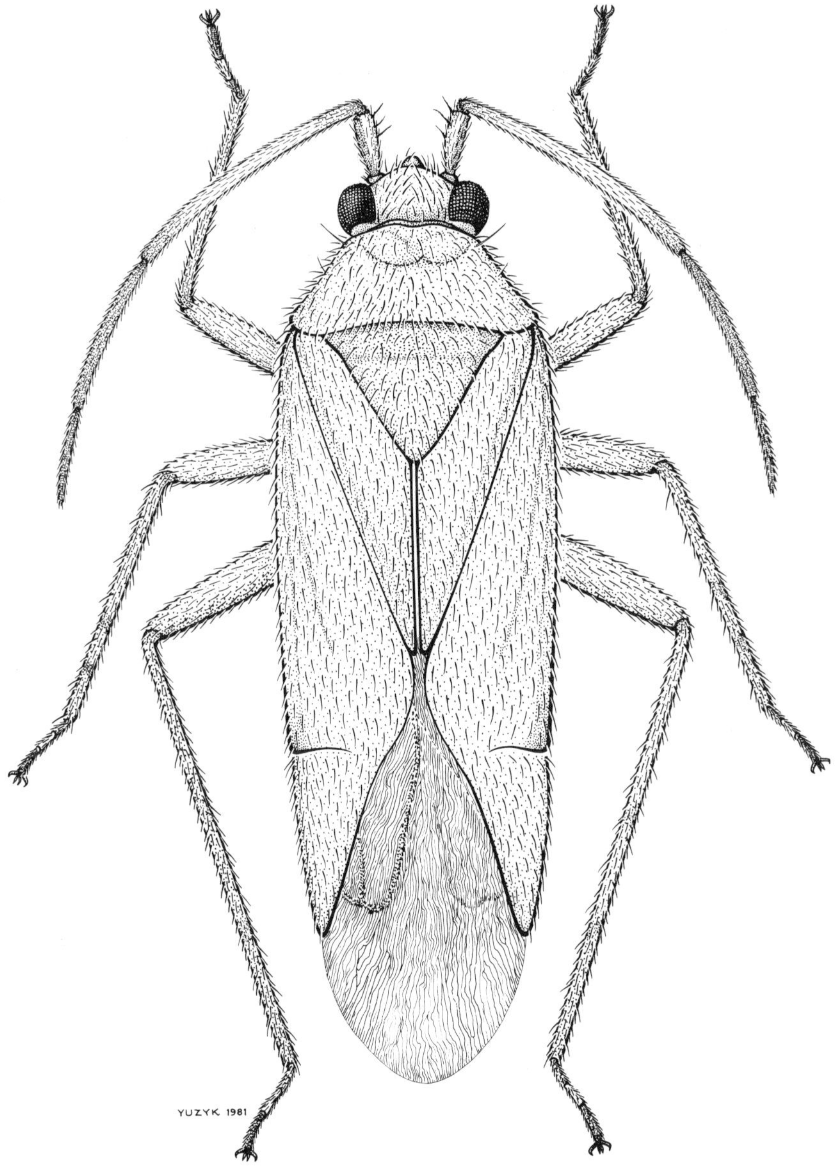

Fig. 86. *Orthotylus nassatus*

Length 4.7–5.0 mm; width 1.5–1.6 mm. Head green; first and second antennal segments green, first segment black on ventral surface, terminal segments light brown. Rostrum 1.2–1.3 mm long, extending to apex of mesosternum. Pronotum and hemelytra green. Veins on wing membrane green.

Remarks. Henry (1977) first reported this European species from Pennsylvania, and Kelton (1982) from Prince Edward Island and Nova Scotia. It is distinguished by the short first antennal segment with few long bristles (Fig. 86), by the black ventral surface on the first antennal segment, and by the left clasper (Fig. 97).

Collected on pear in Nova Scotia; predaceous on pear psylla and aphids.

Also collected on *Acer plantanoides, Tilia cordata,* and *Juglans* cinerea.

Distribution. Europe; Pennsylvania; Prince Edward Island, Nova Scotia (Map 48).

Tribe Pilophorini

The tribe is represented by one genus and two species.

Genus *Pilophorus* Hahn

Antlike, black or brown species. Head oblique, base of head convex, overlapping apex of pronotum. Scutellum tumid with clumps of sericeous pubescence. Hemelytra banded with transverse bars of sericeous pubescence.

Two species, introduced from Europe, were collected. Overwinter in the egg stage.

The nymphs appear about mid-June and the adults about mid-July. The adults are active throughout July and August, and gradually die out by mid-September.

Key to species of *Pilophorus*

1. Head, pronotum, and hemelytra without pilose hairs, only short appressed hairs (Fig. 87) .... *perplexus* **Douglas & Scott** (p. 117)
   Head, pronotum, and hemelytra with pilose hairs, in addition to short hairs (Fig. 88) .... *confusus* **(Kirschbaum)** (p. 118)

116

Pilophorus perplexus Douglas & Scott

Fig. 87; Map 49

Pilophorus perplexus Douglas and Scott, 1875:101.

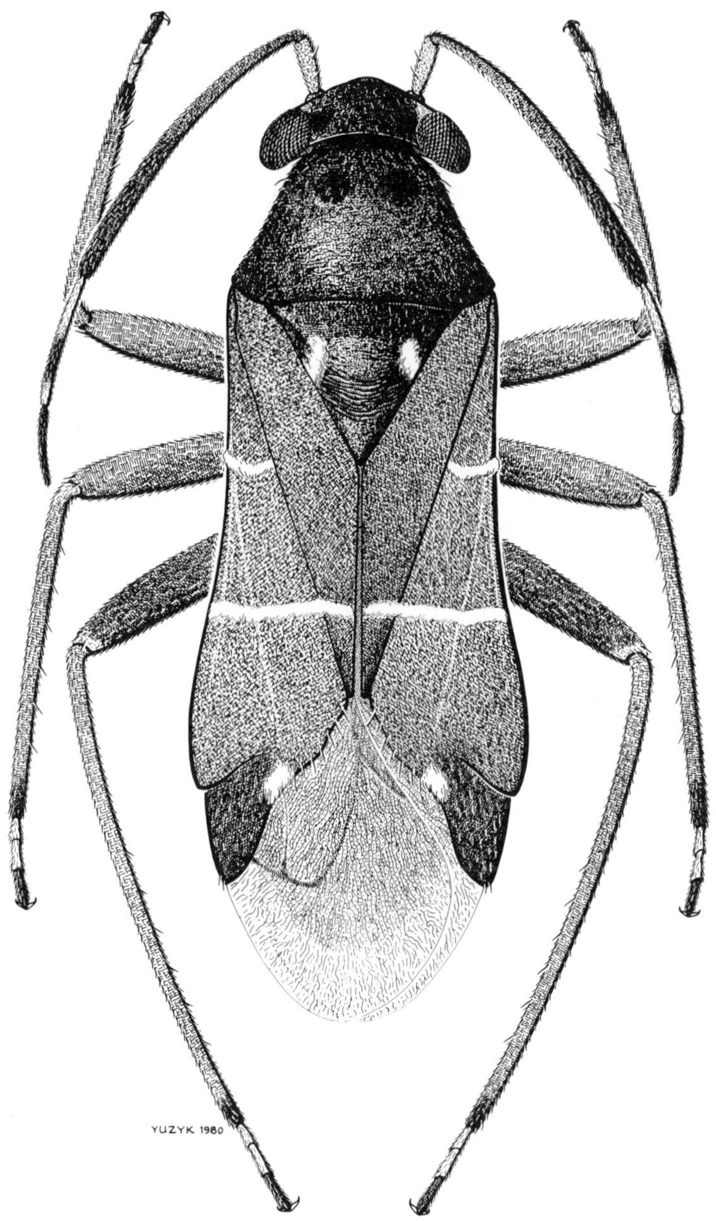

Fig. 87. *Pilophorus perplexus*

Length 4.2–4.4 mm; width 1.4–1.5 mm. Head between eyes black, below eyes brown. Pronotum and scutellum black. Hemelytra pruinose, mostly brown, short; apex of corium with row of long setae; wing membrane with large basal area brown. Ventral surface black; legs brown.

Remarks. Knight (1923*b*) first reported this European species from eastern United States. It is distinguished by the continuous silvery band on the hemelytra, not dislocated at the claval sutures (Fig. 87).

Collected on apple, plum, and pear in Nova Scotia; on apple in Quebec and Ontario; predaceous on mites, aphids, and other small arthropods. Knight (1924) reported the species preying on aphids, Lord (1949) observed the species preying on mites, and MacPhee and Sanford (1954) reported the species preying on mites and codling moth larvae in Nova Scotia.

Distribution. Connecticut, New York; Nova Scotia, Quebec, Ontario (Map 49).

Pilophorus confusus (Kirschbaum)

Fig. 88; Map 50

Capsus confusus Kirschbaum, 1855:293.
Pilophorus confusus: Reuter, 1875*a*:86.

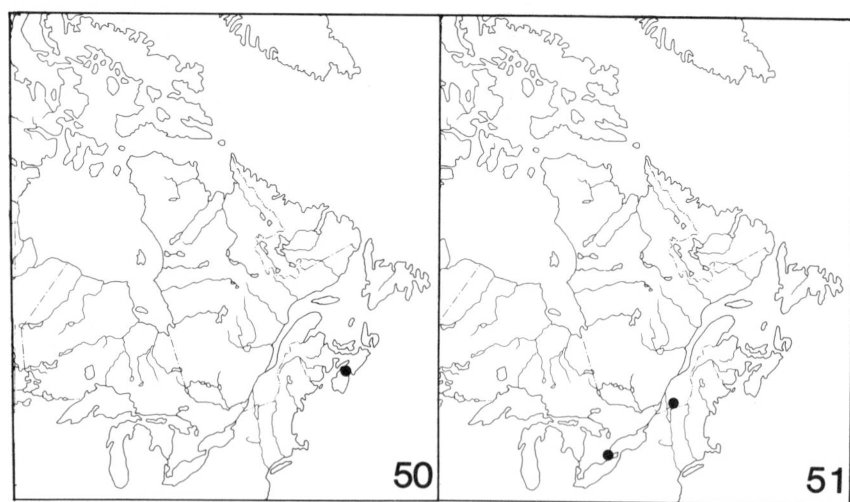

Map 50. Collection locality for *Pilophorus confusus*.
Map 51. Collection localities for *Rhinocapsus vanduzeei*.

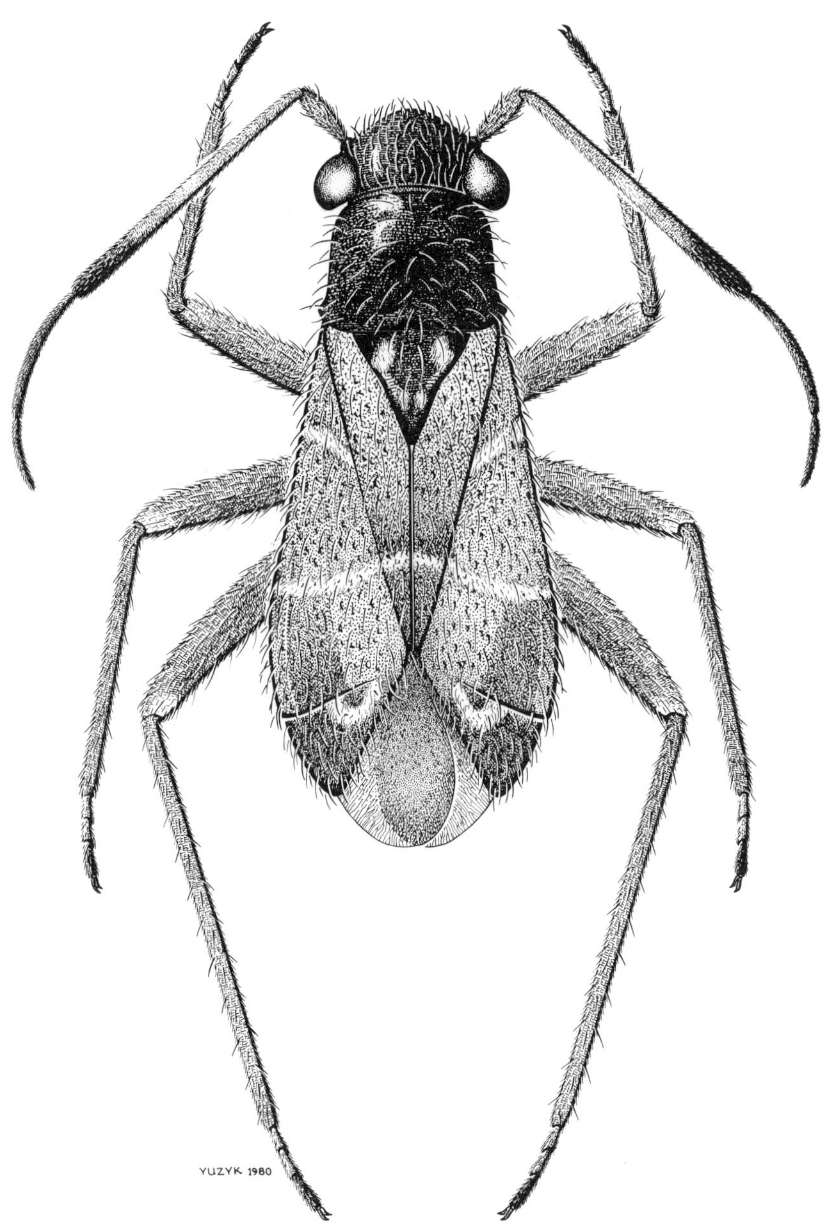

Fig. 88. *Pilophorus confusus*

Length 3.5–3.7 mm; width 1.4–1.5 mm. Head between eyes black, below eyes brown. Pronotum and scutellum black. Hemelytra brown; cuneus and wing membrane reduced; pubescence pilose, intermixed with short, sericeous hairs.

Remarks. Kelton (1982) first reported this European species from Nova Scotia. It is distinguished by the pilose pubescence and by the partly reduced hemelytra (Fig. 88).

Collected on plum in Nova Scotia; predaceous on aphids.

Distribution. Europe; Nova Scotia (Map 50).

Subfamily Phylinae Douglas & Scott

The following are the subfamily characteristics: 1) straight hairlike parempodia between the claws; 2) pulvilli present; 3) male genitalia with rigid ductus seminis, with or without sclerites; and 4) distinctive left clasper.

The subfamily is represented by 1 tribe, Phylini, 7 genera, and 10 species. Three species are predaceous, seven are both predaceous and phytophagous.

Key to genera of Phylinae

1. Hemelytra with one type of pubescence, simple hairs only (Fig. 98) 2
   Hemelytra with two types of pubescence, simple and sericeous hairs (Fig. 104) 4
2. Hind tibia without black spots at bases of spines; dark reddish species (Fig. 98) .... ***Rhinocapsus* Uhler** (p. 122)
   Hind tibia with black spots at bases of spines; black green and black, or green species (Figs. 99–103) .... ***Plagiognathus* Fieber** (p. 122)
4. Second antennal segment strongly swollen (Fig. 104) ....
   .... ***Atractotomus* Fieber** (p. 134)
   Second antennal segment linear 5
5. Hemelytra with white scalelike pubescence intermixed with simple hairs (Fig. 105) .... ***Lepidopsallus* Knight** (p. 136)
   Hemelytra with sericeous hairs intermixed with simple hairs 6
6. Small species 2.8 mm long; hemelytra yellowish green (Fig. 106) ....
   .... ***Campylomma* Reuter** (p. 138)
   Larger species 3.6 mm long; hemelytra spotted with brown (Fig. 107) .....
   .... ***Psallus* Fieber** (p. 140)

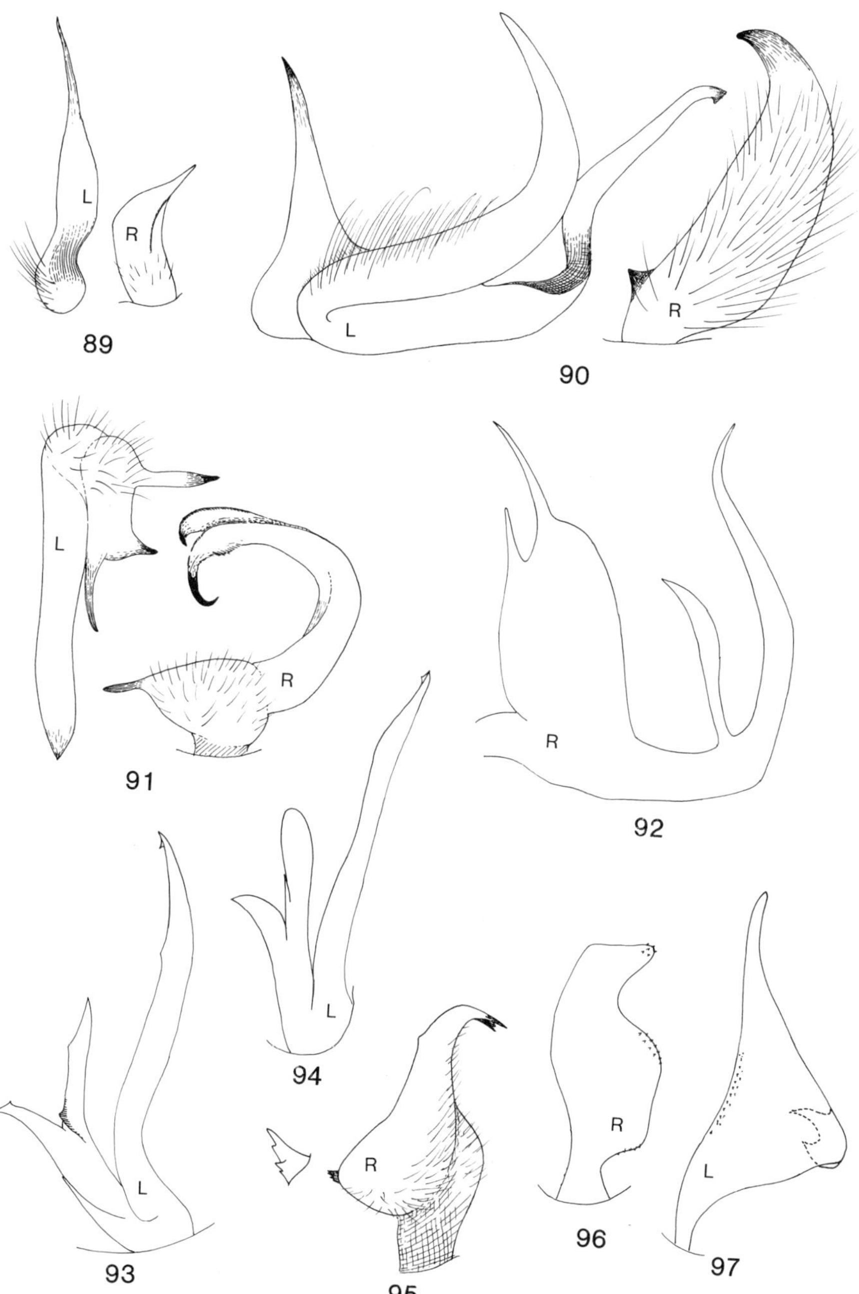

Figs. 89–97. Male claspers of Orthotylini. 89, *Ceratocapsus pilosulus*; 90, *Ceratocapsus modestus*; 91, *Ceratocapsus digitulus*; 92, *Ceratocapsus incisus*; 93, *Ceratocapsus pumilus*; 94, *Ceratocapsus fuscinus*; 95, *Lopidea dakota*; 96, *Orthotylus viridinervis*; 97, *Orthotylus nassatus*.

Genus *Rhinocapsus* Uhler

Elongate-oval, reddish brown species. Head oblique. Pronotum and hemelytra impunctate and shiny; pubescence simple, brown to black. Tibiae without dark spots.

One species was collected. Overwinters in the egg stage.

Rhinocapsus vanduzeei Uhler

Fig. 98; Map 51

Rhinocapsus vanduzeei Uhler, 1890:82.

Length 3.6–3.8 mm; width 1.4–1.6 mm. Head reddish brown, clypeus and surrounding area black. First antennal segment pale, second segment pale on basal half, black on apical half. Hind femur spotted with black.

Remarks. This species is distinguished by the shiny appearance, by the black clypeus, and by the spotted hind femur (Fig. 98).

Collected on raspberry in Quebec and Ontario; predaceous on aphids.

The nymphs appear about mid-May and the adults about mid-June. The adults are active throughout July, and gradually die out by the end of August.

Also collected on *Kalmia polifolia*.

Distribution. Eastern USA; Quebec, Ontario, Manitoba (Map 51).

Genus *Plagiognathus* Fieber

Elongate-oval, black or green, or black and green species. Head oblique. Pronotum and hemelytra shiny, pubescence simple, pale or black. Tibial spines with black spots at bases.

Five species were collected, one introduced from Europe. Overwinter in the egg stage.

The nymphs appear about mid-May and the adults mid-June. The adults are active throughout the summer, and gradually die out by the middle of September.

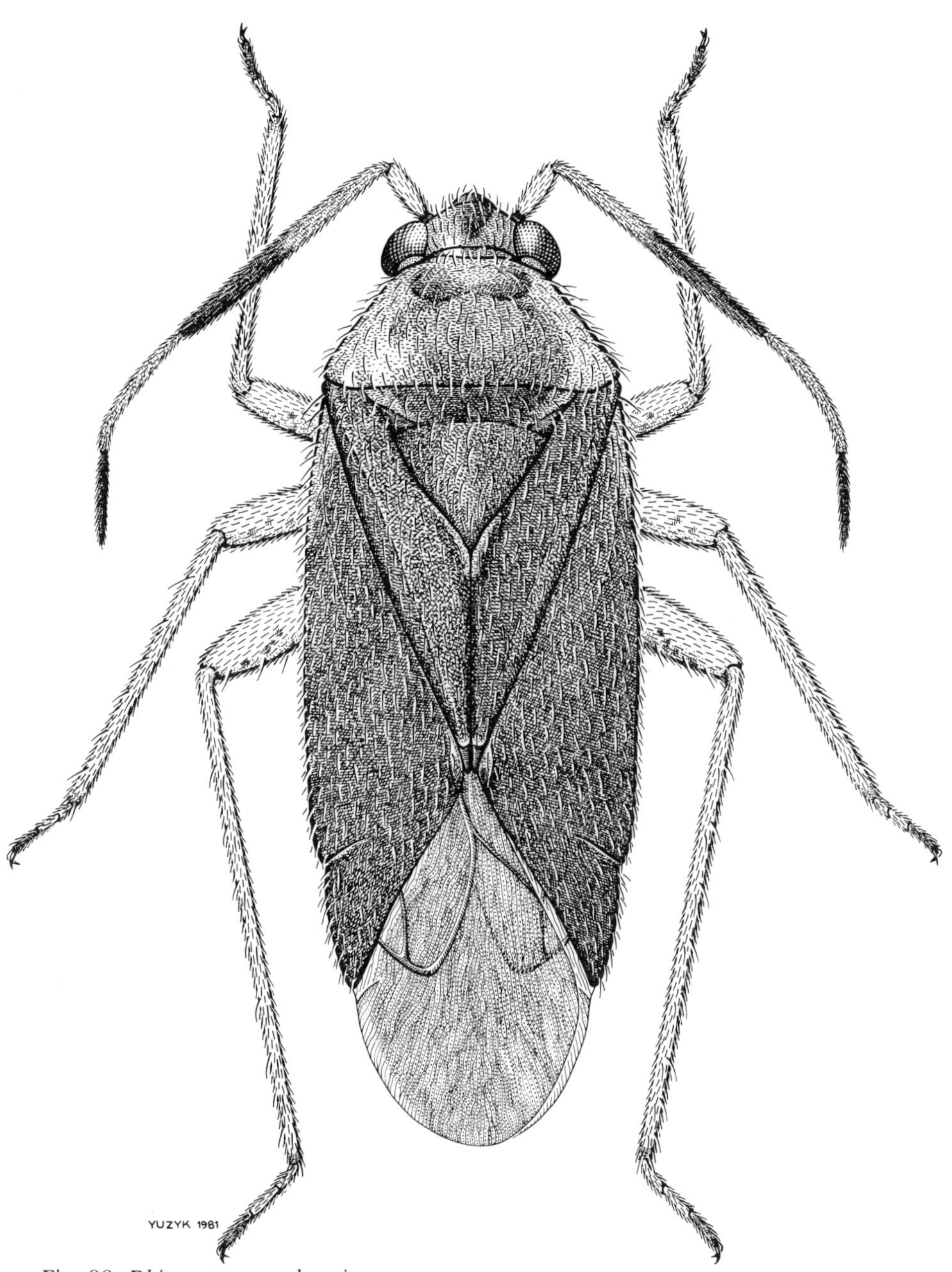

Fig. 98. *Rhinocapsus vanduzeei*

Key to species of *Plagiognathus*

1. Hemelytra black; femora mostly black (Fig. 99) .... ***politus* Uhler** (p. 124)
   Hemelytra pale green, or pale green and black; femora mostly pale green
   2
2. Hemelytra pale green 3
   Hemelytra pale green and black 4
3. Femora with longitudinal black lines on anterior surface; pubescence on dorsum pale (Fig. 100) .... ***ribesi* Kelton** (p. 126)
   Femora without black lines but with black spots; pubescence on dorsum black (Fig. 101) .... ***chrysanthemi* (Wolff)** (p. 126)
4. Cuneus pale green and black (Fig. 100) .... ***obscurus* Uhler** (p. 128)
   Cuneus pale green (Fig. 101) .... ***alboradialis* Knight** (p. 132)

Plagiognathus politus Uhler

Fig. 99; Map 52

Plagiognathus politus Uhler, 1895:52.

Length 3.5–3.8 mm; width 1.3–1.5 mm. Head black, vertex often pale; first and second antennal segments black. Pronotum, scutellum, and hemelytra black; base of cuneus pale green. Ventral surface black; femora black, apices pale green; tibiae pale green, spotted with black.

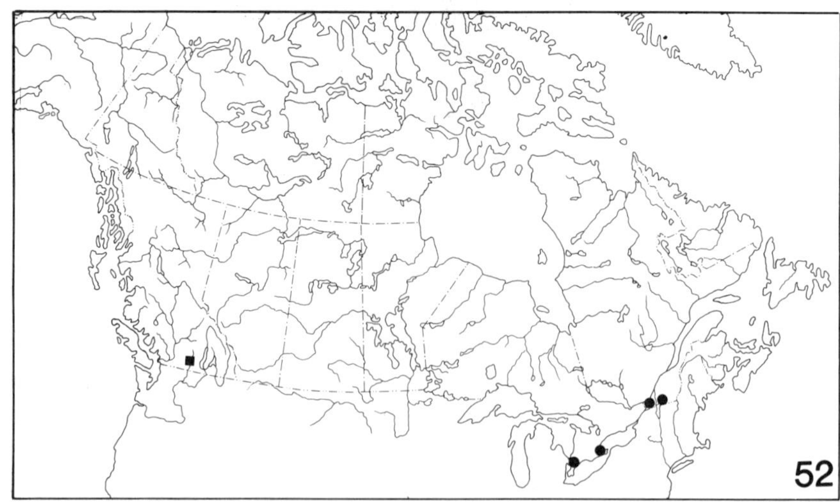

Map 52. Collection localities for *Plagiognathus politus* (●), and *Plagiognathus ribesi* (■).

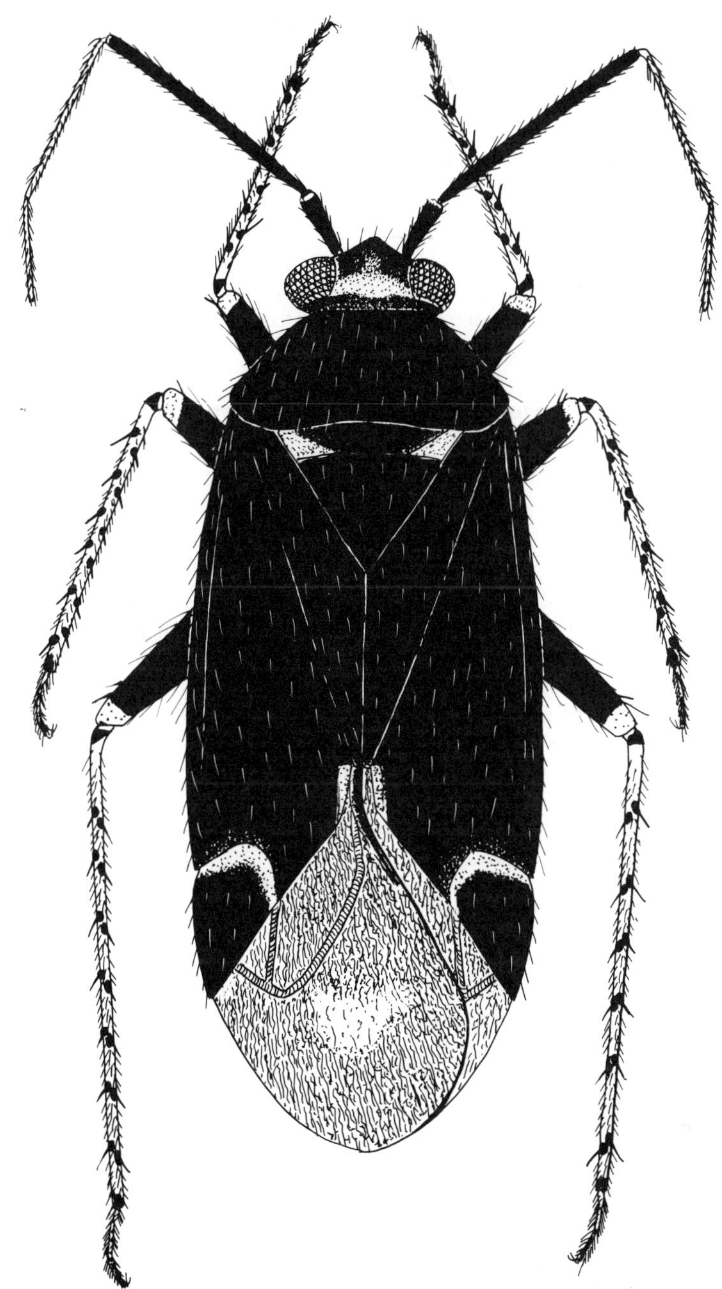

Fig. 99. *Plagiognathus politus*

Remarks. This species is distinguished by the black second antennal segments, by the black hemelytra, and by the black femora with pale green apices (Fig. 99).

Collected on apple and raspberry in Quebec; on apple, pear, peach, mulberry, and raspberry in Ontario; phytophagous, and predaceous on aphids and mites. Knight (1923*b*) reported the species as phytophagous on apple in New York.

Also collected on many other plants.

Distribution. Widespread in USA; Quebec, Ontario (Map 52).

Plagiognathus ribesi Kelton

Fig. 100; Map 52

Plagiognathus ribesi Kelton, 1982:169.

Length 2.8–3.2 mm; width 1.2–1.4 mm. Head yellowish green. First antennal segment pale green, longitudinal line black; second segment pale green, base black. Pronotum, scutellum, and hemelytra pale green; wing membrane with short transverse black bar at apex of cuneus. Legs pale green, each femur with longitudinal black stripe on anterior surface, hind femur with stripe on posterior surface.

Remarks. This species is distinguished by the pale green color, by the black lines on the first antennal segments and femora, and by the transverse black bar on the wing membrane (Fig. 100).

Collected on currant and gooseberry in British Columbia; predaceous on aphids and psyllids.

The nymphs appear about the first part of May and the adults about the first part of June. The adults are active throughout June, July, and August, and gradually die out by early September.

Also collected on *Purshia tridentata* and *Shepherdia canadensis*.

Distribution. Colorado; British Columbia (Map 52).

Plagiognathus chrysanthemi (Wolff)

Fig. 101; Map 53

Miris chrysanthemi Wolff, 1804:157.
Plagiognathus chrysanthemi: Reuter, 1883:452.

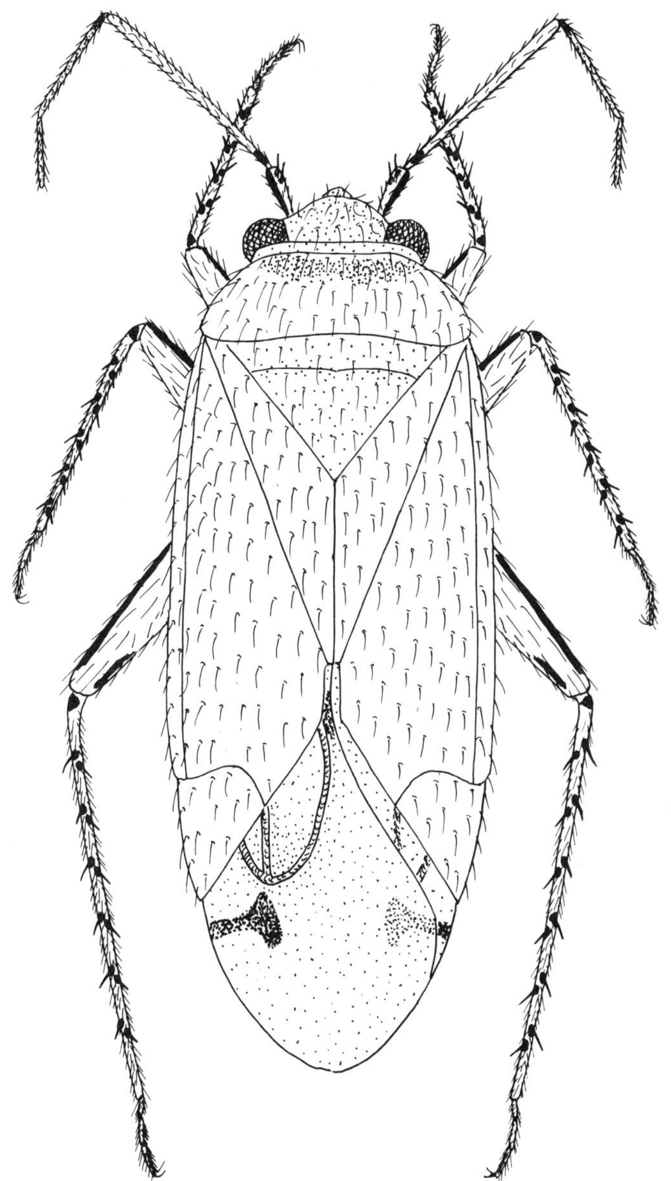

Fig. 100. *Plagiognathus ribesi*

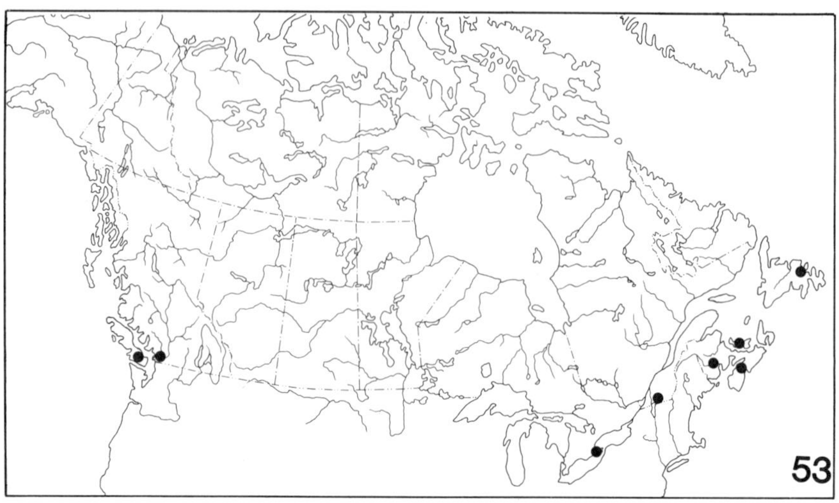

Map 53. Collection localities for *Plagiognathus chrysanthemi*.

Length 3.6–4.2 mm; width 1.3–1.5 mm. Head yellowish brown, tip of clypeus black. First antennal segment green, base and apex black. Pronotum, scutellum, and hemelytra pale green; pubescence black. Legs pale green, femora and tibiae spotted with black.

Remarks.   Knight (1921*b*) first reported this European species from New York, and Tonks (1952) reported it from British Columbia. It is distinguished by the pale green color, by the black pubescence on the dorsum, by the black spotting on the legs, and by the black tip on the clypeus (Fig. 101).

Collected on raspberry, thimbleberry, loganberry, and blackberry in British Columbia; on raspberry and thimbleberry in Ontario, Quebec, New Brunswick, and Nova Scotia; on raspberry in Prince Edward Island and Newfoundland; and on strawberry in Newfoundland; phytophagous, and predaceous on aphids.

Also collected on *Chrysanthemum* spp., and many other herbaceous plants.

Distribution.   Northeastern USA; Atlantic Provinces, Quebec, Ontario, British Columbia (Map 53).

Plagiognathus obscurus Uhler

Fig. 102; Map 54

Plagiognathus obscurus Uhler, 1872:418.
Lygus brunneus Provancher, 1872:104.

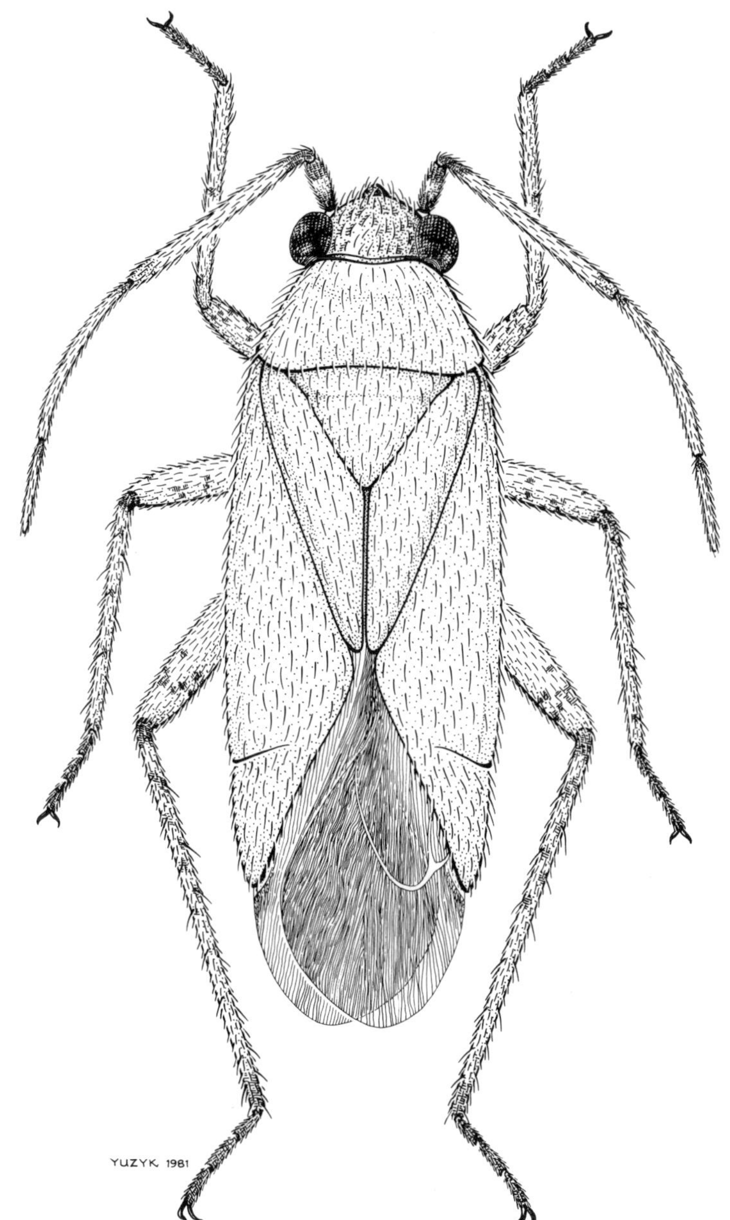

Fig. 101. *Plagiognathus chrysanthemi*

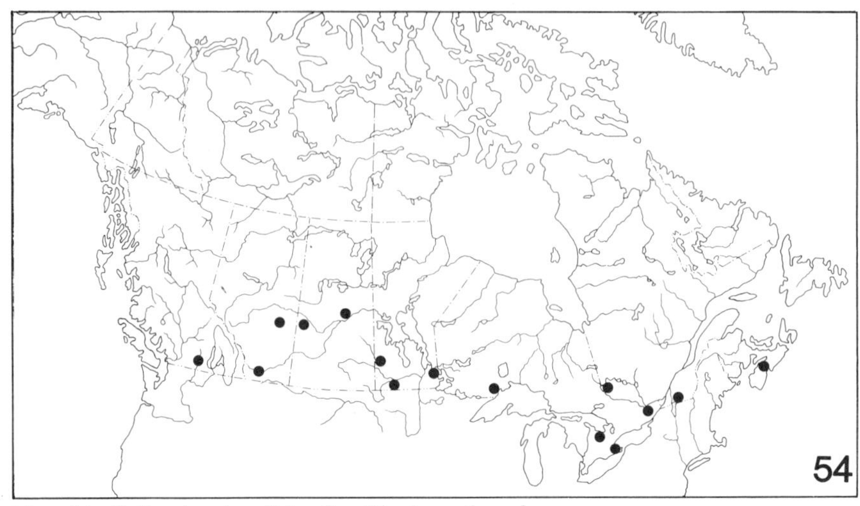

Map 54. Collection localities for *Plagiognathus obscurus*.

Length 4.2–4.6 mm; width 1.5–1.6 mm. Head black, vertex often yellowish; first and second antennal segments black. Pronotum black, often pale yellow at middle. Scutellum black, side margins often pale yellow. Hemelytra black; outer margin of clavus, basal half of corium, and basal half of cuneus pale green. Ventral surface black; legs pale green, hind femur often fuscous.

Remarks. This species is distinguished by the pale green areas on the hemelytra (Fig. 102). It is widely distributed in Canada and is both harmful and beneficial.

Collected on raspberry in Newfoundland; on apple and raspberry in Prince Edward Island; on apple, plum, and raspberry in Nova Scotia; on apple, raspberry, and blueberry in New Brunswick; on apple, raspberry, and thimbleberry in Quebec and Ontario; on raspberry and saskatoon in the Prairie Provinces; and on raspberry, thimbleberry, blackberry, loganberry, and apple in British Columbia; phytophagous and predaceous on aphids and mites. Gilliatt (1935) observed the species on apple preying on aphids and mites; MacPhee and Sanford (1954) reported the species preying on mites, aphids, and codling moth eggs and larvae.

Also collected on a great variety of herbaceous plants.

Distribution. Widespread in USA; British Columbia, Prairie Provinces, Ontario, Quebec, Atlantic Provinces (Map 54).

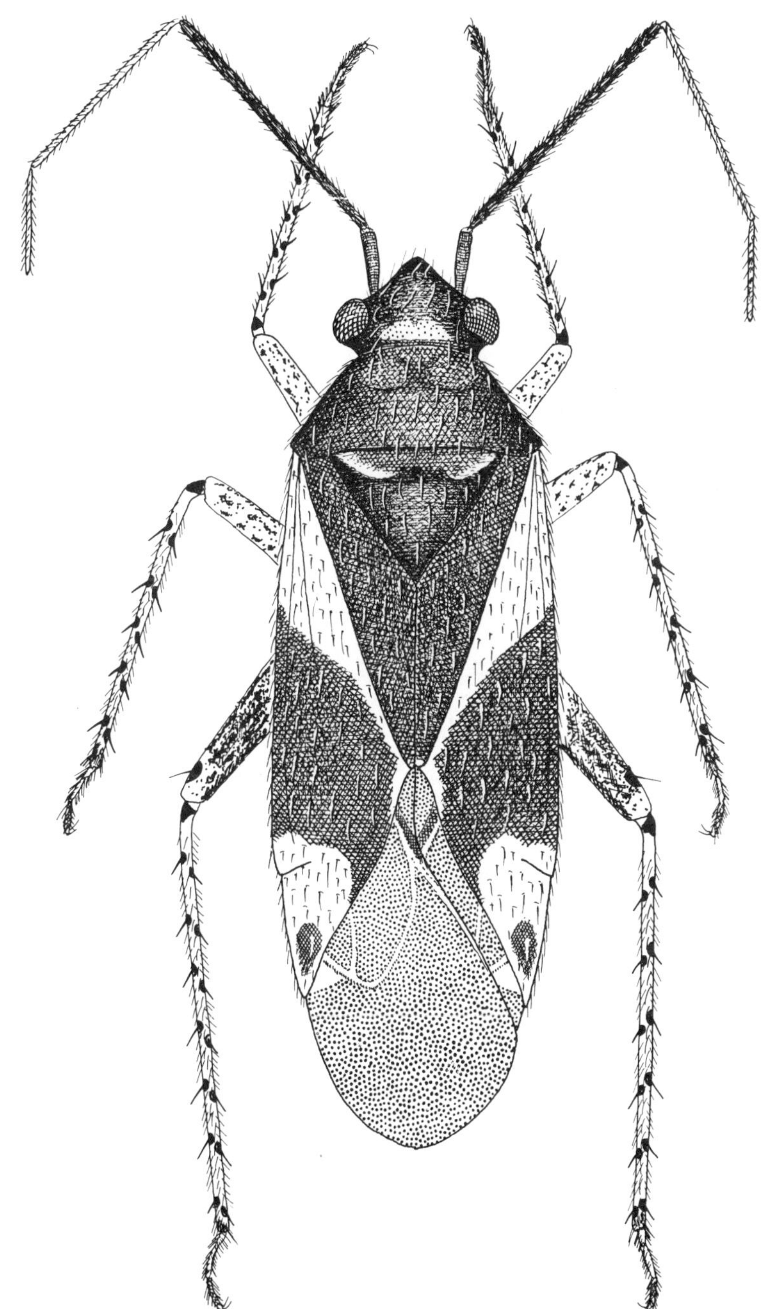

Fig. 102. *Plagiognathus obscurus*

Plagiognathus alboradialis Knight

Fig. 103; Map 55

Plagiognathus alboradialis Knight, 1923*b*:439.

Length 4.5–4.9 mm; width 1.5–1.7 mm. Head black, vertex pale green. First and second antennal segments black. Pronotum black, median area often pale. Scutellum black, side margins often pale green. Hemelytra black, base of corium and all cuneus pale green.

Remarks. This species is distinguished by the elongate form and uniformly pale green cuneus (Fig. 103).

Collected on raspberry in New Brunswick, Quebec, Ontario, and Prairie Provinces; phytophagous, and predaceous on aphids.

Also collected on *Salix* spp.

Distribution. Northeastern USA; New Brunswick, Quebec, Ontario, Prairie Provinces (Map 55).

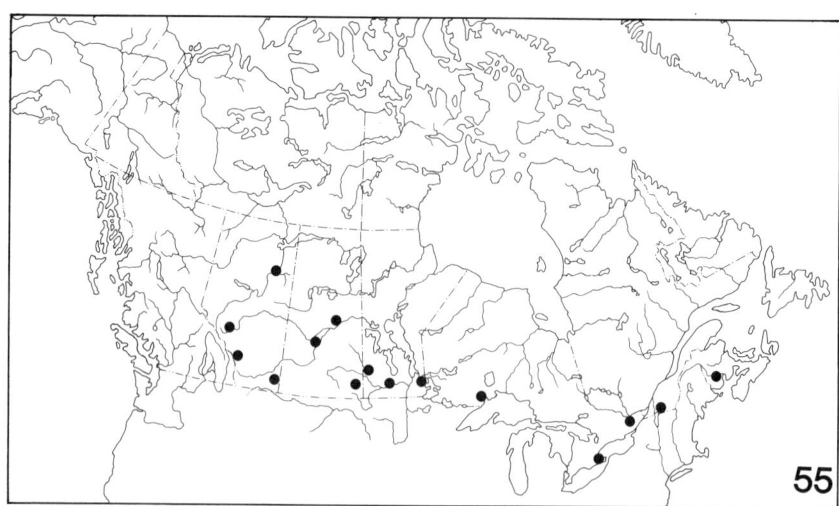

Map 55. Collection localities for *Plagiognathus alboradialis*.

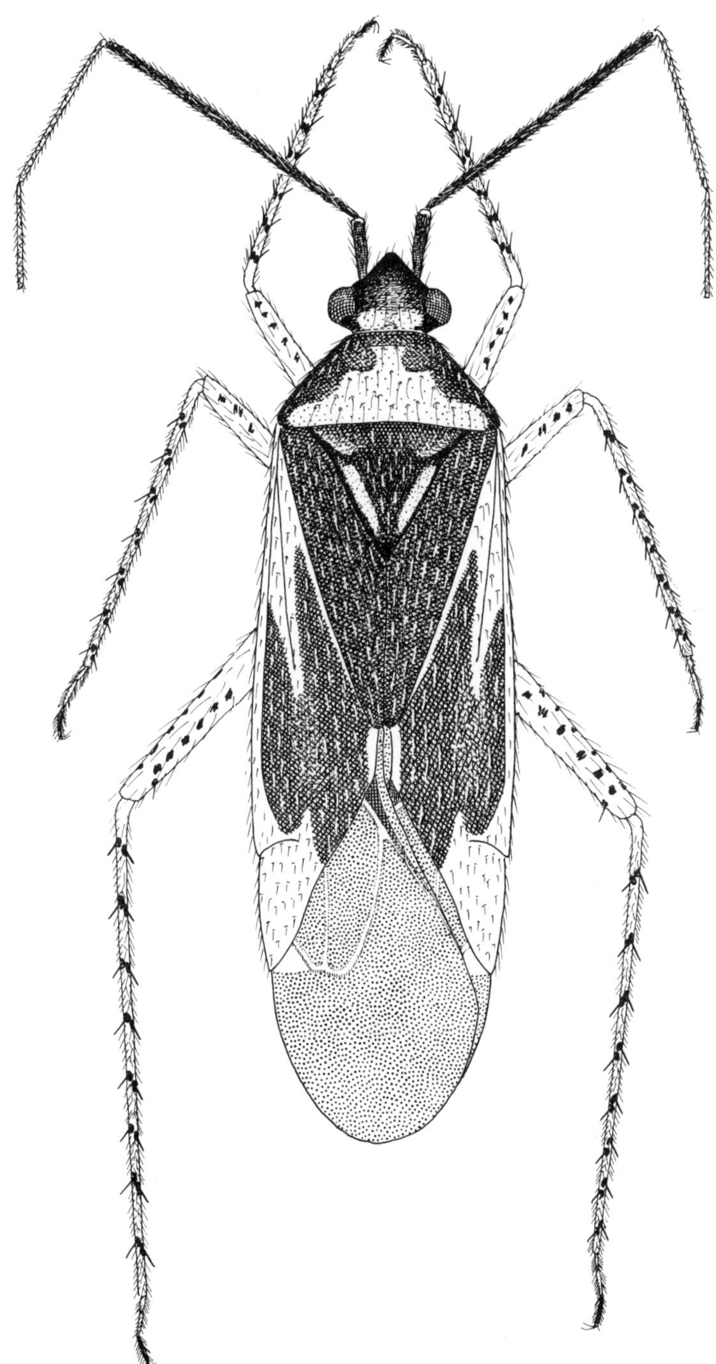

Fig. 103. *Plagiognathus alboradialis*

133

Genus *Atractotomus* Fieber

Oval, black species. Head oblique, base truncate. Pronotum and hemelytra finely rugose; pubescence scalelike, intermixed with black or golden simple hairs.

One species, introduced from Europe, was collected. Overwinters in the egg stage.

Atractotomus mali (Meyer)

Fig. 104; Map 56

Capsus mali Meyer, 1843:63.
Atractotomus mali: Fieber, 1861:296.

Length 3.0–3.5 mm; width 1.4–1.6. First and second antennal segments black, second segment greatly inflated. Dorsum and ventral surface with white, scalelike hairs. Femora black, tibiae pale green with black spines.

Remarks. Knight (1924) first reported this European species from Nova Scotia. It is distinguished by the inflated second antennal segment and by the scalelike pubescence (Fig. 104).

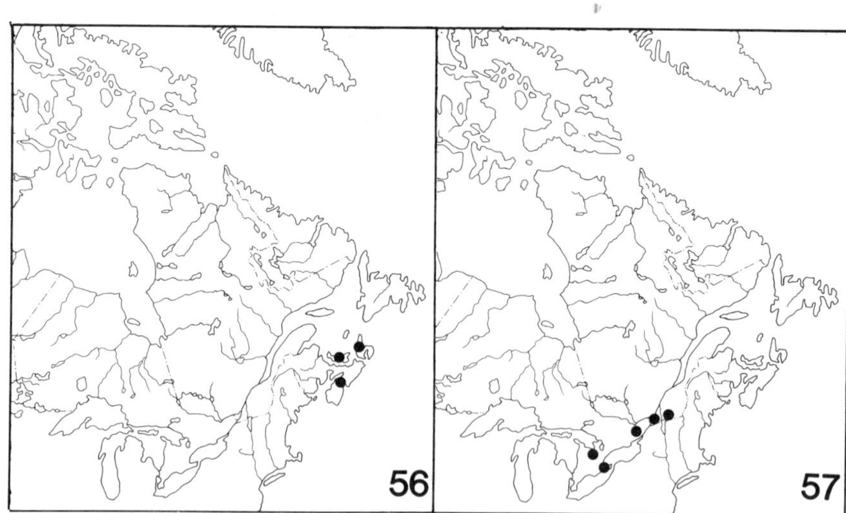

Map 56. Collection localities for *Atractotomus mali*.

Map 57. Collection localities for *Lepidopsallus minisculus*.

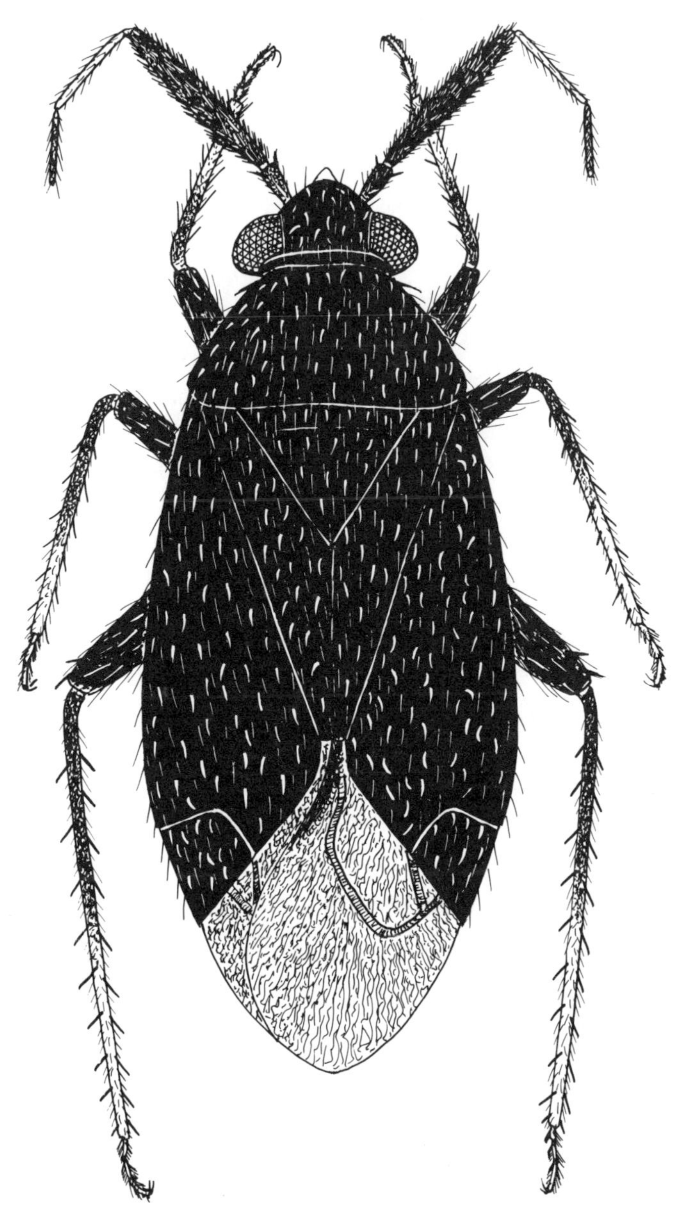

Fig. 104. *Atractotomus mali*

Collected on apple in Prince Edward Island; on apple, pear, and plum in Nova Scotia; phytophagous, and predaceous on aphids and mites. Lord (1949) and MacPhee and Sanford (1954, 1956) referred to this species as *Criocoris saliens* and considered it as phytophagous and predaceous.

The nymphs appear about the end of May and the adults about the end of June. The adults are active throughout July and August, and gradually die out by the end of August.

Also collected on *Crataegus* spp.

Distribution.   Europe; Prince Edward Island, Nova Scotia (Map 56.)

Genus *Lepidopsallus* Knight

Ovate, black species. Head oblique, base truncate. Pronotum and hemelytra finely rugose; pubescence scaly, intermixed with simple black hairs. Legs black.

One species was collected. Overwinters in the egg stage.

Lepidopsallus minisculus Knight

Fig. 105; Map 57

Lepidopsallus minisculus Knight, 1923b:472.

Length 2.6–2.8 mm; width 1.3–1.4 mm. Second antennal segment shorter than head width. Dorsal and ventral surface black.

Remarks.   This species is distinguished by the short second antennal segment and by the scaly pubescence (Fig. 105).

Collected on apple in Quebec and Ontario; predaceous on mites and aphids. Knight (1923b) observed the species on apple in New York.

The nymphs appear about mid-May and the adults about mid-June. The adults are active throughout July and August, and gradually die out by the end of August.

Also collected on *Crataegus chrysocarpa*.

Distribution.   New York; Quebec, Ontario (Map 57).

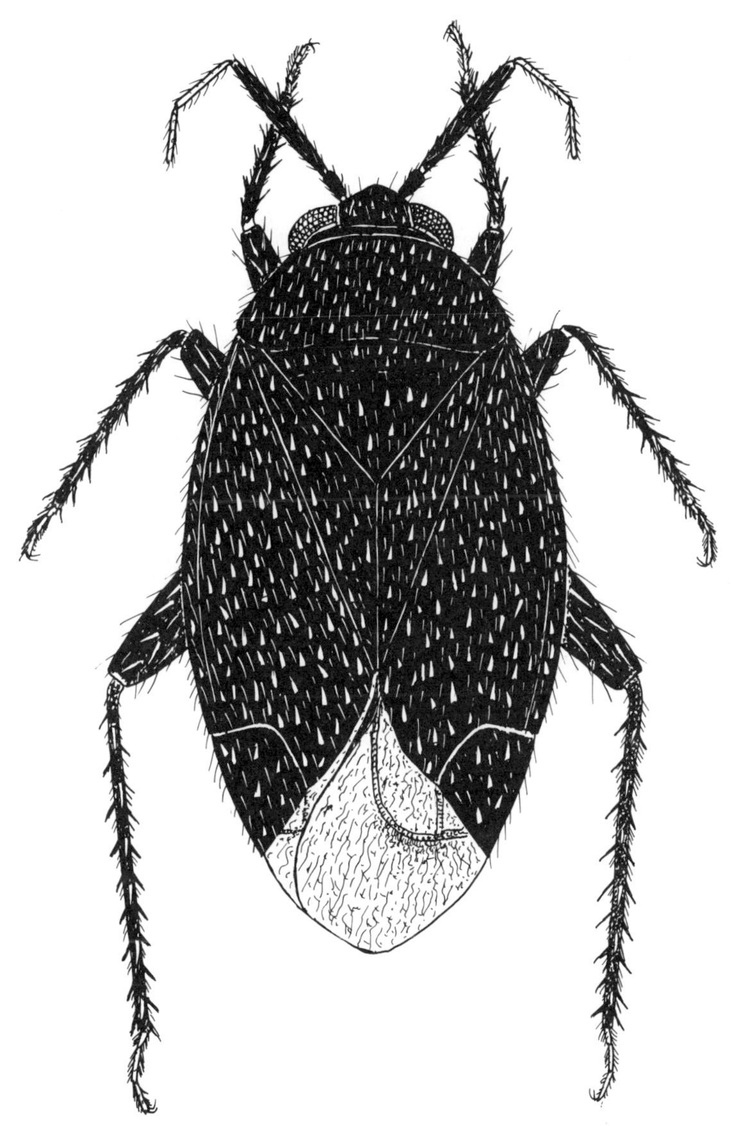

Fig. 105. *Lepidopsallus minisculus*

Genus *Campylomma* Reuter

Small, pale yellowish green species. Head nearly vertical, base truncate. Pronotum and hemelytra finely punctate; pubescence sericeous, intermixed with simple hairs. Ventral surface black; legs green, spotted with black.

One species, introduced from Europe, was collected. Overwinters in the egg stage.

Campylomma verbasci (Meyer)

Fig. 106; Map 58

Capsus verbasci Meyer, 1843:70.
Campylomma verbasci: Reuter, 1878:53.

Length 2.6–2.8 mm; width 1.1–1.3 mm. Head yellow, clypeus often black. Antennae green, apex of first segment and base of second segment black; second segment shorter than head width. Pronotum and hemelytra yellowish green. Legs pale green, spotted with black.

Remarks.   Uhler (1886) first recorded this European species from eastern United States, and Provancher (1887) from Ontario. It is distinguished by the small size, by the markings on the antennae, and by the spotting on the legs (Fig. 106).

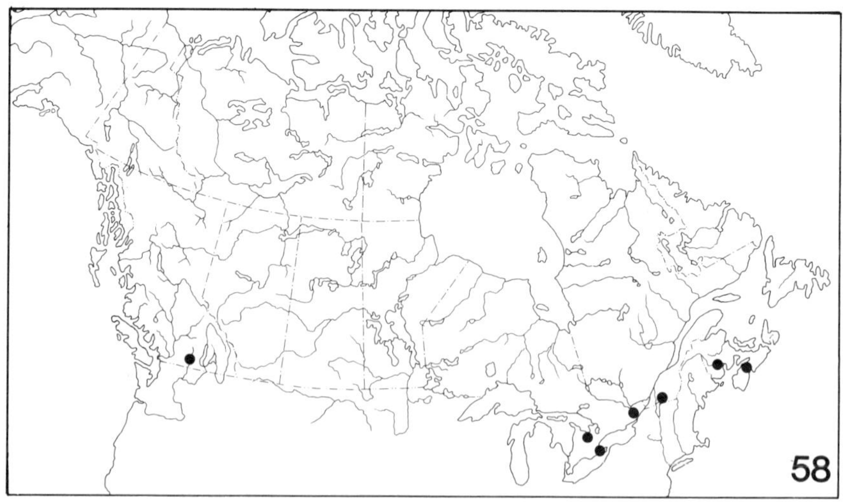

Map 58. Collection localities for *Campylomma verbasci*.

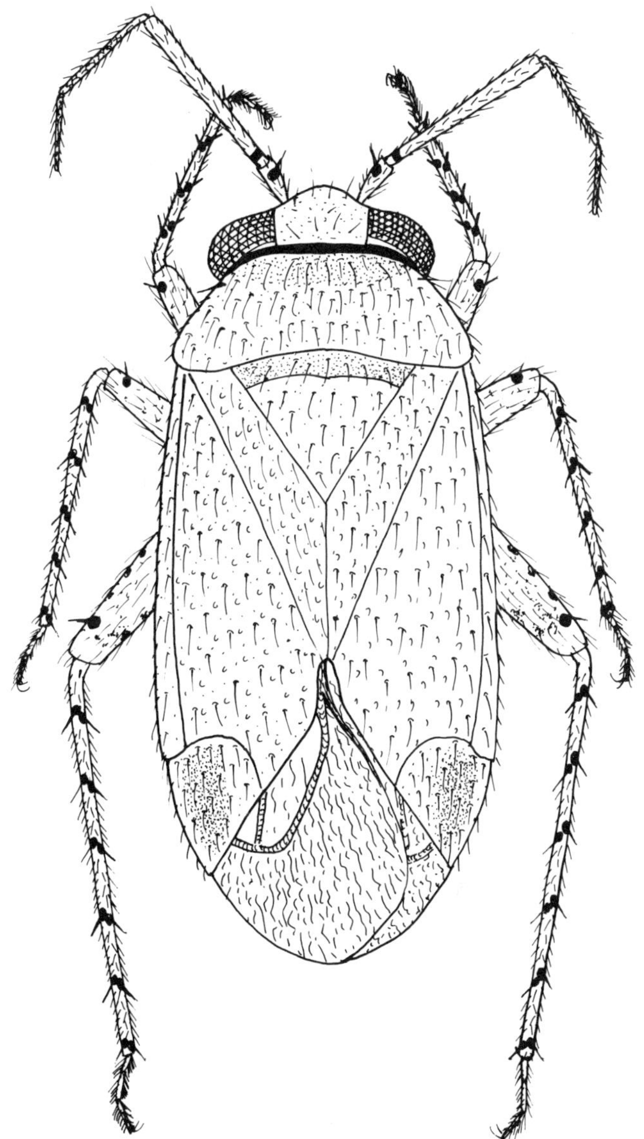

Fig. 106. *Campylomma verbasci*

Collected on apple in Nova Scotia, New Brunswick, and Quebec; on apple, pear, and peach in Ontario; on apple and pear in British Columbia; phytophagous, and predaceous on aphids, psyllids, and mites. Ross and Caesar (1919) first observed the species on apple in Ontario; Gilliatt (1930) observed the species on apple in Nova Scotia; Tonks (1952), McMullen and Jong (1967), and Madsen and Morgan (1975) collected the species in British Columbia on loganberry, pear, and grape, respectively. Gilliatt (1935) reported that there may be two generations per year.

The nymphs appear about mid-May and the adults about mid-June. The adults are active throughout July and August, and gradually die out in September.

Also breeds on *Verbascum thapsus* and many other herbaceous plants.

Distribution.   Northeastern and northwestern USA; Nova Scotia, New Brunswick, Quebec, Ontario, British Columbia (Map 58).

Genus *Psallus* Fieber

Elongate-oval, green species. Head oblique, second antennal segment longer than head width. Pronotum and hemelytra impunctate; pubescence sericeous, intermixed with simple hairs. Femora spotted with black, tibiae with black spots at bases of spines.

One species, introduced from Europe, was collected. Overwinters in the egg stage.

Psallus salicellus (Herrich-Schaeffer)

Fig. 107; Map 59

Capsus salicellus Herrich-Schaeffer, 1841:47.
Psallus salicellus: Fieber, 1861:305.
Coniortodes salicellus: Sanford and Herbert, 1966:997.
Psallus salicellus: Kelton, 1982:172.

Length 3.6–3.9 mm; width 1.3–1.4 mm. Head yellow; first antennal segment pale green, spot at base and apex fuscous; second segment pale green. Pronotum and scutellum yellowish green, often spotted with fuscous. Hemelytra pale green; clavus and corium spotted with fuscous; inner margin of cuneus often with red mark at middle. Ventral surface fuscous, hind femur often heavily infuscated.

Remarks.   MacPhee and Sanford (1961) first reported this European species from Nova Scotia as *Psallus* sp., and Sanford and Herbert (1966) as *Coniortodes salicellus*. Kelton (1982) also reported it from Prince Edward Island and British Columbia, and referred to it as *Psallus salicellus*.

140

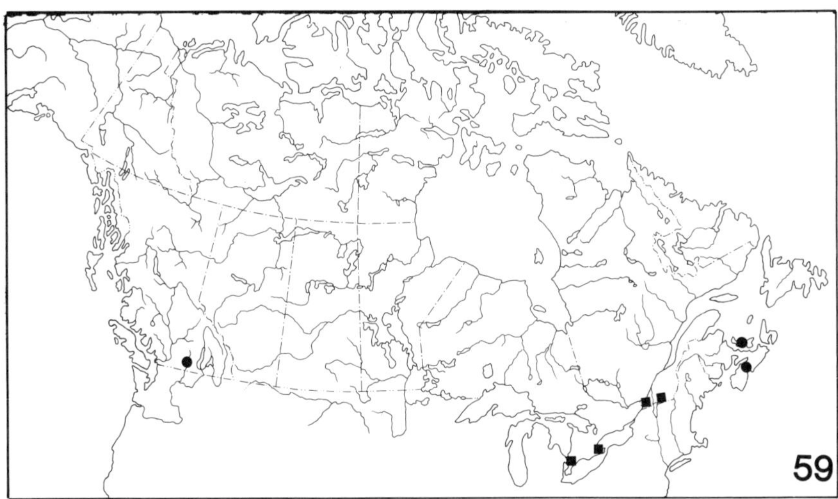

Map 59. Collection localities for *Psallus salicellus* (●), and *Hyaliodes vitripennis* (■).

It is distinguished by the pale green color and by the fuscous spotting on the pronotum and hemelytra (Fig. 107).

Collected on raspberry in Prince Edward Island; on apple in Nova Scotia; on raspberry and thimbleberry in British Columbia; predaceous on aphids. Sanford and Herbert (1966) observed the species preying on mites in Nova Scotia.

Also collected on *Corylus avellana*.

The nymphs appear about mid-June and the adults about mid-July. The adults are active throughout August, and gradually die out in September.

Distribution.   Europe; Prince Edward Island, Nova Scotia, British Columbia (Map 59).

Subfamily Deraeocorinae Douglas & Scott

The following are the subfamily characteristics: 1) straight hairlike parempodia between the claws; 2) pulvilli absent; 3) pronotal collar distinct; and 4) pronotum deeply punctate.

The subfamily is represented by 2 tribes, 3 genera, and 10 species. The species are predaceous.

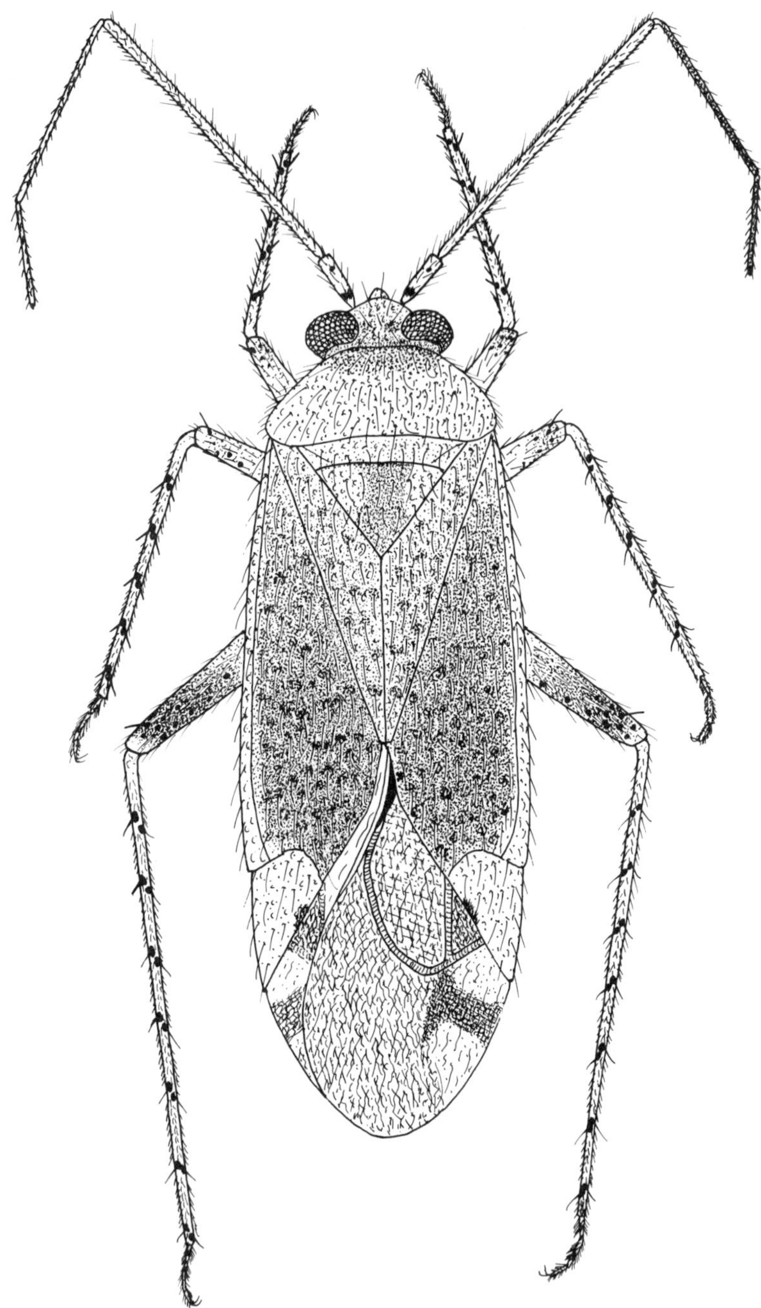

Fig. 107. *Psallus salicellus*

Key to tribes of Deraeocorinae

1. Eyes situated forward from posterior margin of head; wing membrane with one cell (Fig. 15); slender species (Figs. 108,109) .... **Hyaliodini** (p. 143)
   Eyes situated near posterior margin of head; wing membrane with two cells (Fig. 16); broad species (Fig. 110) .... **Deraeocorini** (p. 147)

Tribe Hyaliodini

Two species belonging to the genus *Hyaliodes* were collected.

Genus *Hyaliodes* Reuter

Elongate, shiny species. Head vertical, neck distinct; eyes large, carina between them absent. Pronotum deeply punctate, hemelytra finely punctate; wing membrane with one cell. Legs long, slender.

Two species were collected. Overwinter in the egg stage.

The nymphs appear about the first part of June and the adults about the first part of July. The adults are active throughout July and August, and gradually die out by mid-September.

Key to species of *Hyaliodes*

1. Collar and calli pale green; scutellum mostly pale green; apical margin of corium black (Fig. 108) .... *vitripennis* **(Say)** (p. 143)
   Collar and calli black; scutellum black at base; apical margin of corium red (Fig. 109) .... *harti* **Knight** (p. 145)

Hyaliodes vitripennis (Say)

Figs. 15, 108; Map 59

Capsus vitripennis Say, 1832:24.
Hyaliodes vitripennis: Riley, 1870:137.

Length 4.6–4.9 mm; width 1.5–1.8 mm. Head pale green, neck often black. Rostrum 1.0–1.1 mm long. First antennal segment almost as long as length of pronotum. Pronotum pale green, basal area adjacent to scutellum often black. Scutellum mostly pale green. Hemelytra mostly clear, inner margin of clavus and apical margin of corium black. Legs pale green.

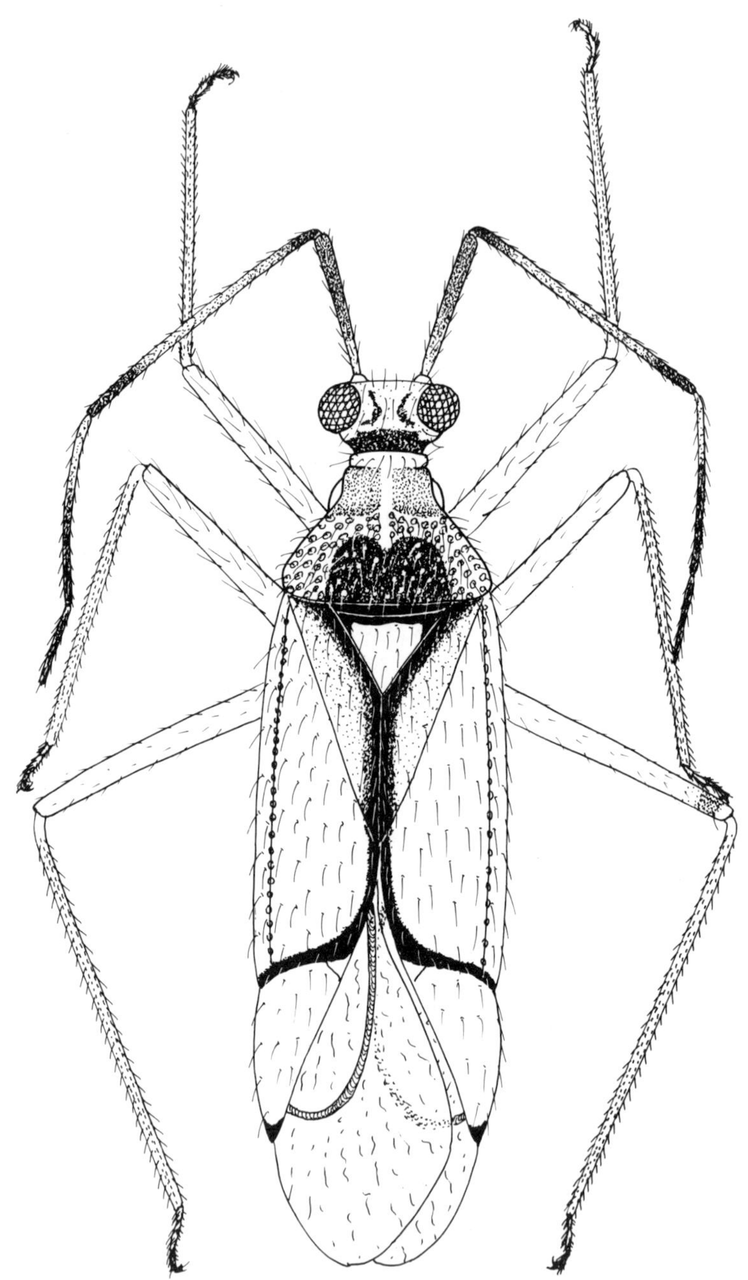

Fig. 108. *Hyaliodes vitripennis*

Remarks. This species is distinguished by the pale green collar, calli, and scutellum and by the black margin on the inner clavus and apical corium (Fig. 108).

Collected on apple and pear in Quebec; on apple, pear, plum, and grape in Ontario; predaceous on mites and aphids.

Also collected on *Quercus* spp., and *Ulmus americana*.

Distribution. Eastern USA; Quebec, Ontario (Map 59).

Hyaliodes harti Knight

Fig. 109; Map 60

Hyaliodes harti Knight, 1941*b*:57.

Length 5.0–5.3 mm; width 1.5–1.8 mm. Head pale green tinged with red, area between eyes dark brown. Rostrum 1.2–1.3 mm long. First antennal segment longer than length of pronotum. Pronotum mostly pale green, collar and calli black. Scutellum mostly black, apex pale. Hemelytra mostly clear; inner margin of clavus black, apical margin of corium red. Legs pale green.

Remarks. Gilliatt (1935) first observed *harti* (cited as *vitripennis* Say) in Nova Scotia. It is distinguished by the black collar and calli and by the red margin on the apical corium (Fig. 107).

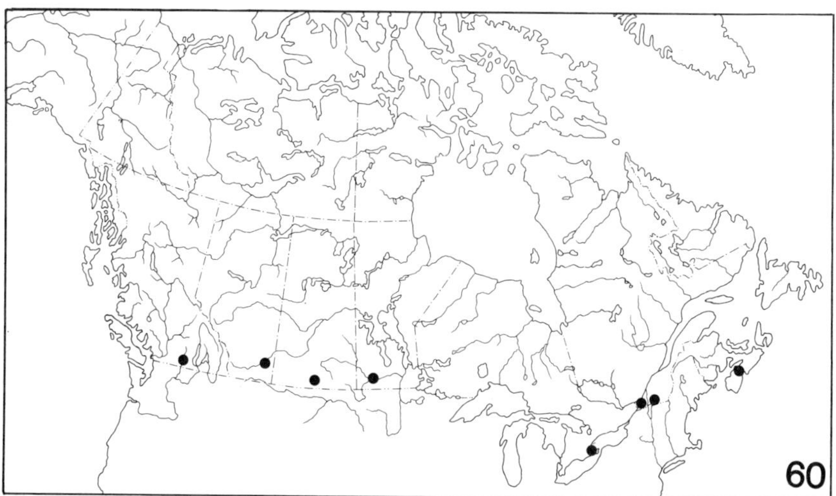

Map 60. Collection localities for *Hyaliodes harti*.

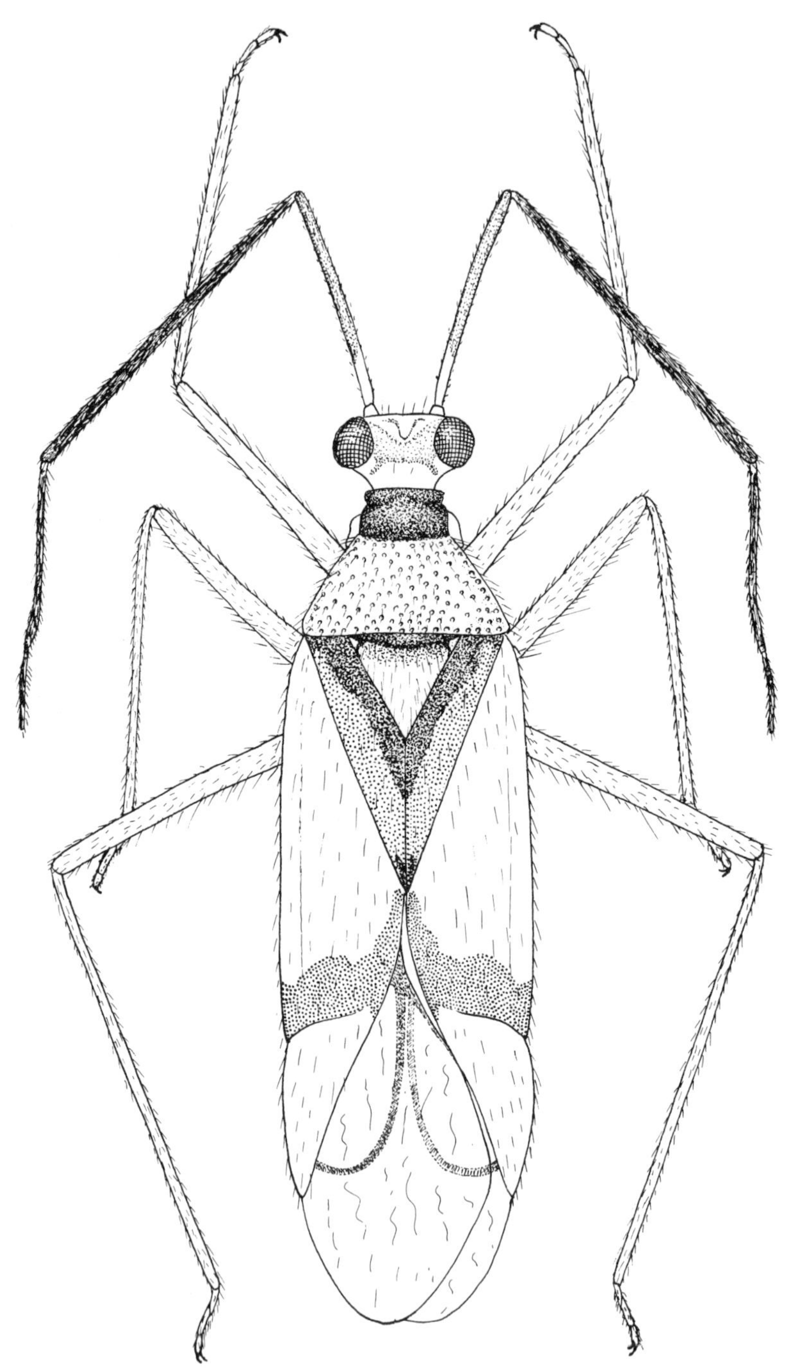

Fig. 109. *Hyaliodes harti*

Collected on apple and plum in Nova Scotia; on apple and pear in Quebec; on apple, pear, peach, plum, apricot, sweet cherry, sour cherry, mulberry, grape, and raspberry in Ontario; on raspberry in the Prairie Provinces; on apple, pear, peach, plum, and sweet cherry in British Columbia; predaceous on mites, aphids, psyllids, and other small arthropods.

Also collected on *Acer* spp., *Alnus rugosa*, *Corylus americana*, *Fraxinus* spp., *Juglans nigra*, *Ostrya virginiana*, *Quercus* spp., *Robinea pseudoacacia*, *Shepherdia canadensis*, *Ulmus americana*, and potato.

Distribution. Eastern and north central USA; Nova Scotia, Quebec, Ontario, Prairie Provinces, British Columbia (Map 60).

Tribe Deraeocorini

Eight species representing the genera *Eurychilopterella* and *Deraeocoris* were collected.

Key to genera of Deraeocorini

1. Head strongly protruded (Fig. 110); rostrum extending beyond hind coxae
. . . . *Eurychilopterella* **Reuter** (p. 147)
Head not strongly protruded (Figs. 111–117); rostrum extending to middle coxa . . . . *Deraeocoris* **Kirschbaum** (p. 148)

Genus *Eurychilopterella* Reuter

Elongate-oval species. Head protruded, carina between eyes distinct. Rostrum extending beyond hind coxae. Pronotum strongly punctate; hemelytra finely punctate; pubescence simple, long and dense, silvery.

One species was collected. Overwinters in the egg stage.

Eurychilopterella luridula Reuter

Figs. 16, 110; Map 61

Eurychilopterella luridula Reuter, 1909:60.

Length 3.9–4.5 mm; width 1.7–2.0 mm. Head black, area next to eyes pale. Rostrum 2.6–2.8 mm long. Pronotum, scutellum, and hemelytra pale green marked with black. Ventral surface mostly black; legs mostly pale green.

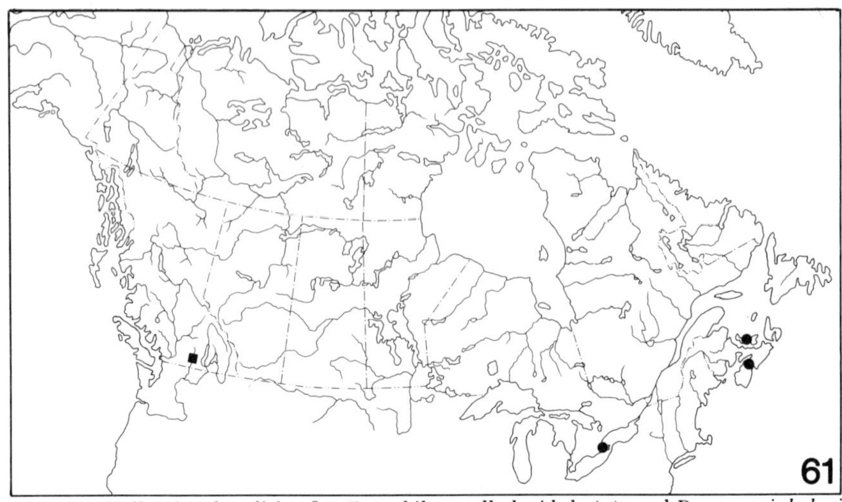

Map 61. Collection localities for *Eurychilopterella luridula* (●), and *Deraeocoris bakeri* (■).

Remarks.   This species is distinguished by the protruding head, by the color pattern on the dorsum, by the long and dense pubescence (Fig. 110), and by the long rostrum.

Collected on apple in Prince Edward Island; on apple and pear in Nova Scotia and Ontario; predaceous on scale insects. MacPhee (*in litt.*) observed large number of nymphs preying on apple scale in Nova Scotia.

The nymphs appear about the end of May and the adults about the end of June. The adults are active throughout July and August, and gradually die out by the end of September.

Also collected on *Quercus rubra* and *Carya ovata*.

Distribution.   Eastern USA; Prince Edward Island, Nova Scotia, Ontario (Map 61).

Genus *Deraeocoris* Kirschbaum

Robust, strongly punctate species. Head oblique, carina between eyes distinct or absent. Rostrum extending to middle coxae. Pronotum and hemelytra strongly punctate; scutellum punctate or impunctate; pubescence simple, short, and sparse.

Seven species were collected. Species with punctate scutellum hibernate; species with impunctate scutellum overwinter in the egg stage.

148

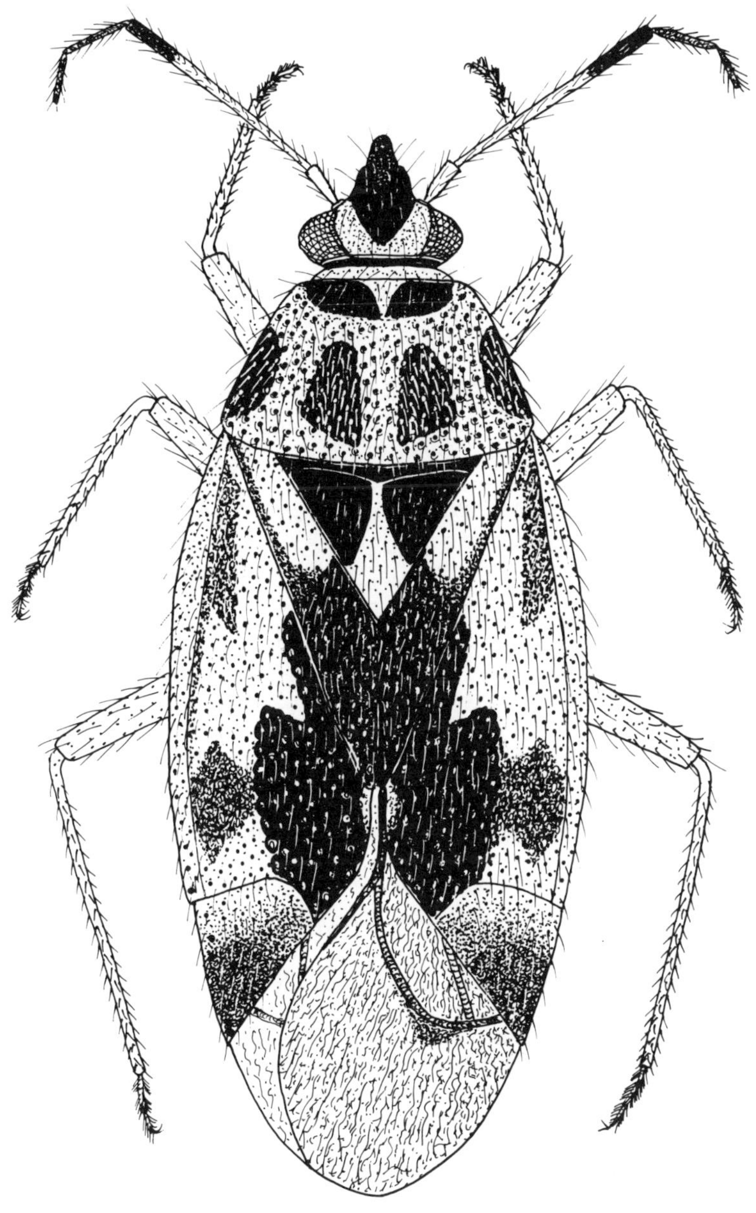

Fig. 110. *Eurychilopterella luridula*

Key to species of *Deraeocoris*

1. Scutellum punctate (Figs. 111–113) 2
   Scutellum impunctate 4
2. Pronotum and hemelytra uniformly black; wing membrane with apical half
   clear (Fig. 111) .... ***bakeri* Knight** (p. 150)
   Pronotum and hemelytra marked with pale areas; wing membrane with apical
   half fuscous or with fuscous spots (Figs. 112, 113) 3
3. Calli distinctly punctate; wing membrane on apical half mostly clear; species
   3.9 mm or smaller (Fig. 112) .... ***nebulosus* (Uhler)** (p. 152)
   Calli impunctate; wing membrane on apical half fuscous; species 4.6 mm or
   larger (Fig. 113) .... ***brevis* (Uhler)** (p. 154)
4. Hind tibiae uniformly pale yellow; apical half of wing membrane clear (Fig.
   114); male claspers (Fig. 125) .... ***nitenatus* Knight** (p. 156)
   Hind tibiae banded with brown; apical half of wing membrane fuscous (Figs.
   115–117) 5
5. Pronotum and hemelytra mostly grayish (Fig. 115); male claspers (Fig. 124)
   .... ***aphidiphagus* Knight** (p. 156)
   Pronotum and hemelytra mostly brown; wing membrane with large fuscous
   spot near apex (Fig. 116) 6
6. Pronotum without black rays behind calli (Fig. 116); male claspers (Fig. 127)
   .... ***fasciolus* Knight** (p. 158)
   Pronotum with black rays behind calli (Fig. 117); male claspers (Fig. 128) ...
   .... ***borealis* (Van Duzee)** (p. 160)

Deraeocoris bakeri Knight

Fig. 111; Map 61

Deraeocoris (Camptobrochis) bakeri Knight, 1921a:102.

Length 3.6–4.2 mm; width 1.7–2.0 mm. Head black, tip of clypeus
and carina often pale yellow. Pronotum black, calli impunctate. Scutellum
black, apex often pale yellow; punctate. Hemelytra black; apical half of
wing membrane clear.

Remarks. This species is distinguished by the small size, by the
black pronotum and hemelytra, and by the clear apical half of the wing
membrane (Fig. 111).

Collected on apple and pear in British Columbia; predaceous on
aphids and psyllids.

The adults hibernate. The eggs are laid in the spring and during
early summer, and the overwintered adults gradually die out by the end of
July. The nymphs appear about the end of April or early May, and new
generation adults appear about the end of May or early June. Thus, the
new adults overlap the declining numbers of overwintered adults. The
new adults are active throughout the summer and fall until hibernation.

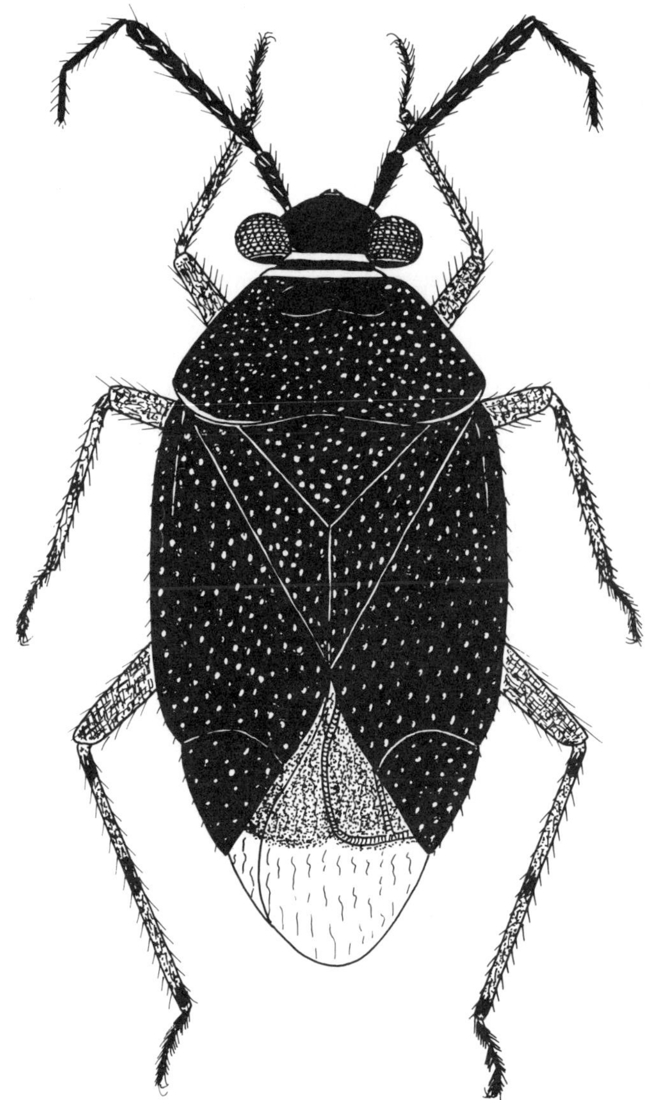

Fig. 111. *Deraeocoris bakeri*

Also collected on *Chrysothamnus nauseosus*, *Artemisia tridentata*, and *Purshia tridentata*, and carrot.

Distribution.   Western USA; British Columbia (Map 61).

Deraeocoris nebulosus (Uhler)

Fig. 112; Map 62

Camptobrochis nebulosus Uhler, 1872:414.
Deraeocoris (Camptobrochis) nebulosus: Knight, 1921a:91.

Length 3.2–4.0 mm; width 1.4–1.9 mm. Head light brown marked with black. Pronotum light brown, callus and spot behind black; calli punctate. Scutellum black, side margins and apex pale yellow; punctate. Hemelytra light brown marked with black; wing membrane mostly clear, spot near outer margin fuscous.

Remarks. This species is distinguished by the small size, by the color pattern on the dorsum, by the punctate calli and scutellum, and by the fuscous spot on the wing membrane (Fig. 112).

Collected on apple and pear in Nova Scotia and Quebec; on apple, pear, peach, sour cherry, and mulberry in Ontario; predaceous on aphids and mites. Gilliatt (1935) observed the species preying on red mites in Nova Scotia; MacPhee and Sanford (1956) observed it preying on mites, aphids, and larvae of eyespotted bud moth and codling moth.

The adults hibernate. The life history is similar to that of *bakeri*, but the nymphs appear in June and the adults in July.

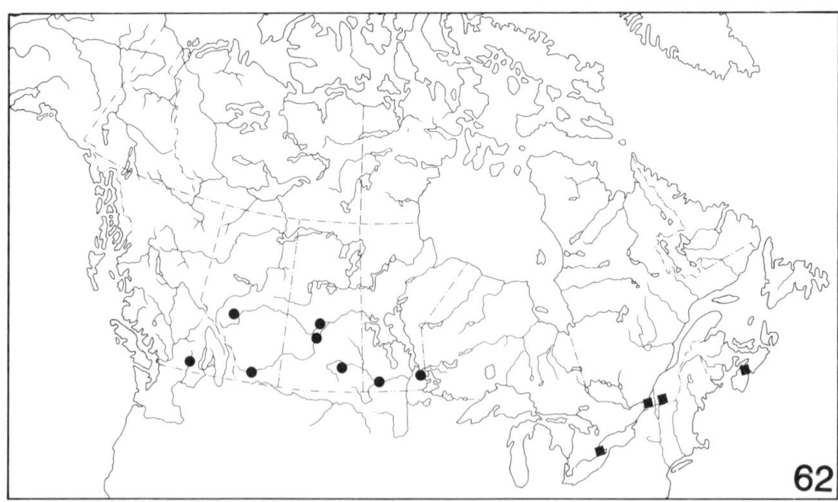

Map 62. Collection localities for *Deraeocoris nebulosus* (■), and *Deraeocoris brevis* (●).

Also collected on *Tilia cordata, Salix* spp., *Quercus* spp., *Juglans nigra*, and *Corylus americana*.

Distribution. Widespread in USA; Nova Scotia, Quebec, Ontario (Map 62).

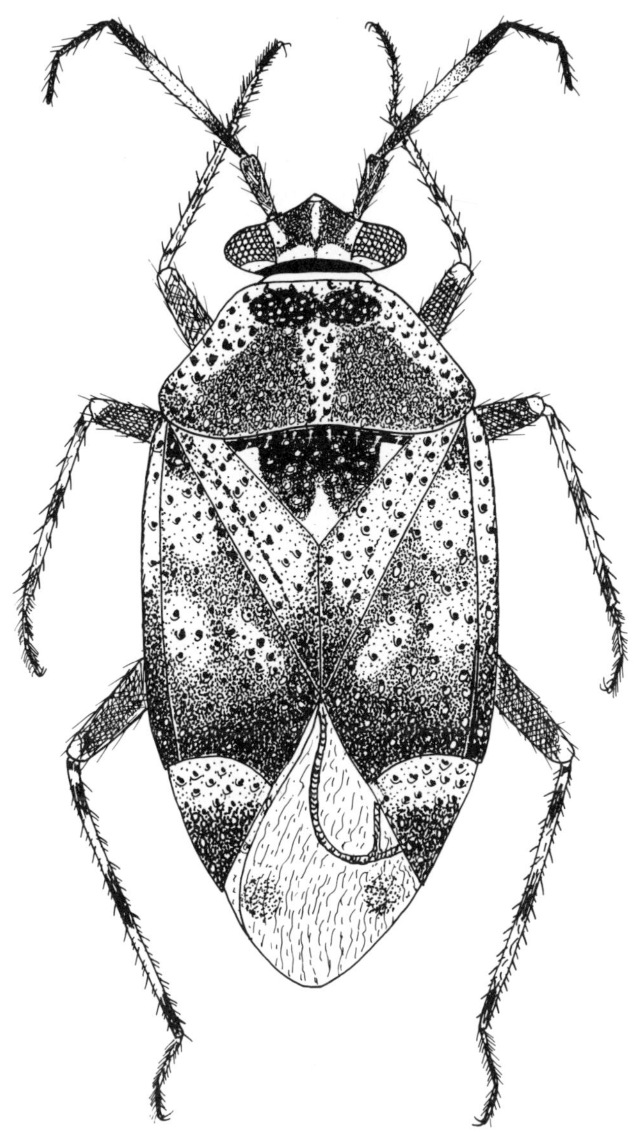

Fig. 112. *Deraeocoris nebulosus*

Deraeocoris brevis (Uhler)

Fig. 113; Map 62

Camptobrochis brevis Uhler, 1904:359.
Deraeocoris (Camptobrochis) brevis: Knight, 1921a:103.

Length 4.6–4.9 mm; width 2.1–2.5 mm. Head black marked with yellow. Pronotum black, side margins often pale brown, callus impunctate. Scutellum black, side margins and apex yellow; punctate. Hemelytra light brown marked with black; wing membrane mostly pale on basal half, fuscous on apical half.

Remarks. This species is distinguished by the color pattern on the dorsum, by the impunctate calli, by the punctate scutellum, and by the fuscous apical half of the wing membrane (Fig. 113).

Collected on apple, pear, and raspberry in British Columbia; on saskatoon, chokecherry, and raspberry in the Prairie Provinces; predaceous on mites, aphids, and psyllids. McMullen and Jong (1967) observed the species preying on mites and pear psylla in British Columbia.

The adults hibernate. The life history is similar to that of *bakeri*.

Also collected on *Artemisia tridentata, Acer negundo, Ceanothus sanguineus, Spiraea douglassi, Pinus contorta, P. banksiana, Juniperus communis*, and carrot.

Distribution. Western USA; British Columbia, Prairie Provinces (Map 62).

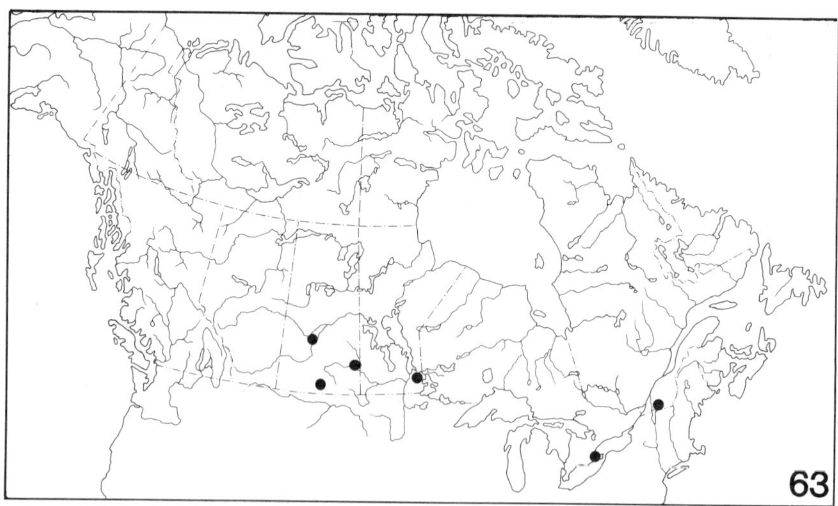

Map 63. Collection localities for *Deraeocoris nitenatus*.

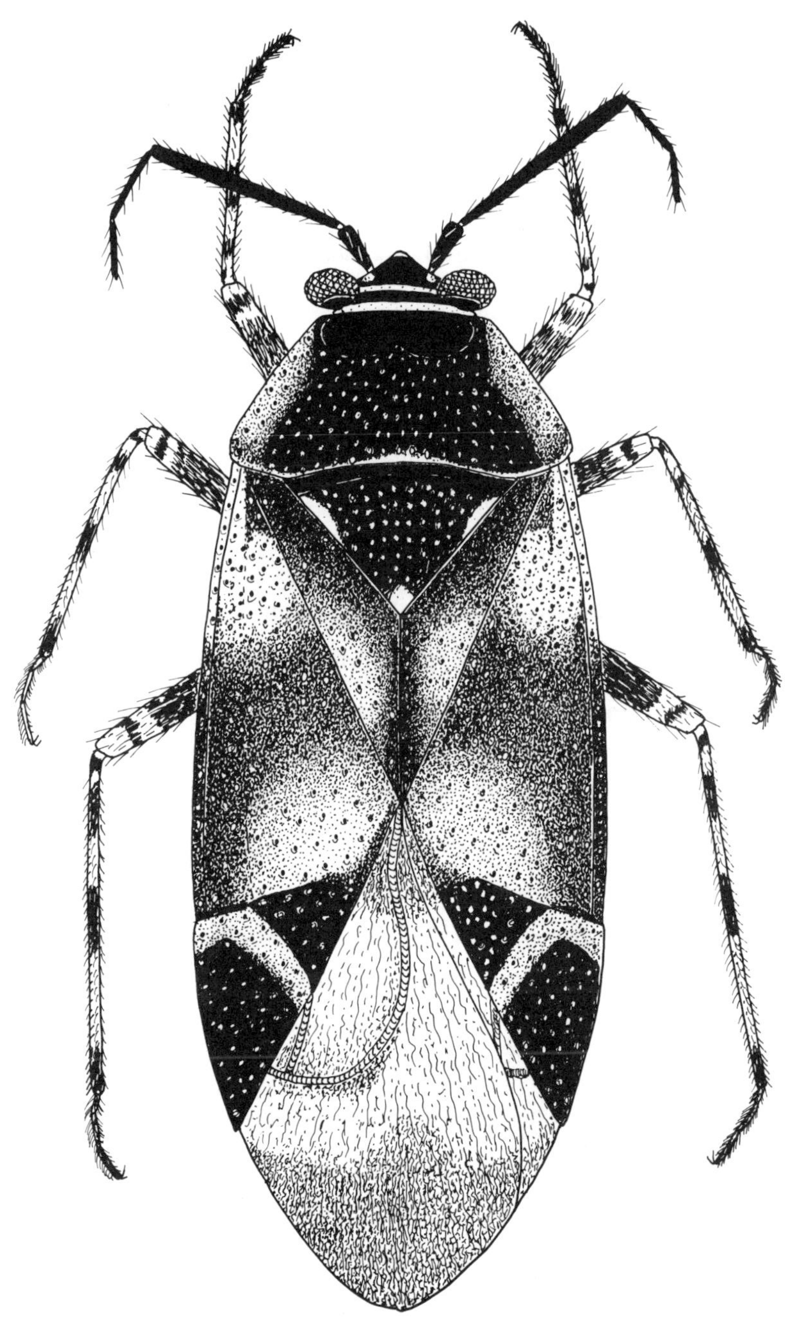

Fig. 113. *Deraeocoris brevis*

Deraeocoris nitenatus Knight

Figs. 114, 125; Map 63

Deraeocoris nitenatus Knight, 1921a:141.

Length 5.6–6.3 mm; width 2.6–2.9 mm. Head mostly pale yellow, frons with several transverse black bars. Pronotum light brown, punctures black; calli margined with black. Scutellum impunctate. Hemelytra light brown; corium often reddish, punctures black. Apical half of wing membrane clear. Tibiae pale yellow.

Remarks.   The species is distinguished by the pale tibiae (Fig. 114) and by the male claspers (Fig. 125).

Collected on chokecherry in Quebec; on apple, pear, and mulberry in Ontario; on chokecherry and saskatoon in the Prairie Provinces; predaceous on aphids and psyllids.

Overwinters in the egg stage. The nymphs appear about the first of May and the adults about early June. The adults are active throughout June, July, and August, and gradually die out by September.

Also collected on *Quercus* spp., and *Ulmus americana*.

Distribution.   Eastern and north central USA; Quebec, Ontario, Prairie Provinces (Map 63).

Deraeocoris aphidiphagus Knight

Figs. 115, 126; Map 64

Deraeocoris aphidiphagus Knight, 1921a:134.

Length 5.7–6.6 mm; width 2.8–3.2 mm. Head pale yellow marked with black. Pronotum gray, calli and punctures black. Scutellum pale with longitudinal bar each side of middle black; impunctate. Hemelytra grayish marked with black. Ventral surface brown; legs mostly pale; hind femur brown on apical half; tibiae banded with brown.

Remarks.   This species is distinguished by the grayish color of the hemelytra (Fig. 115) and by the left clasper (Fig. 126).

Collected on apple in Prince Edward Island, New Brunswick, Quebec, and Ontario; on pear in Nova Scotia; on chokecherry and saskatoon in the Prairie Provinces; predaceous on aphids.

Overwinters in the egg stage. The life history is similar to that of *nitenatus*.

156

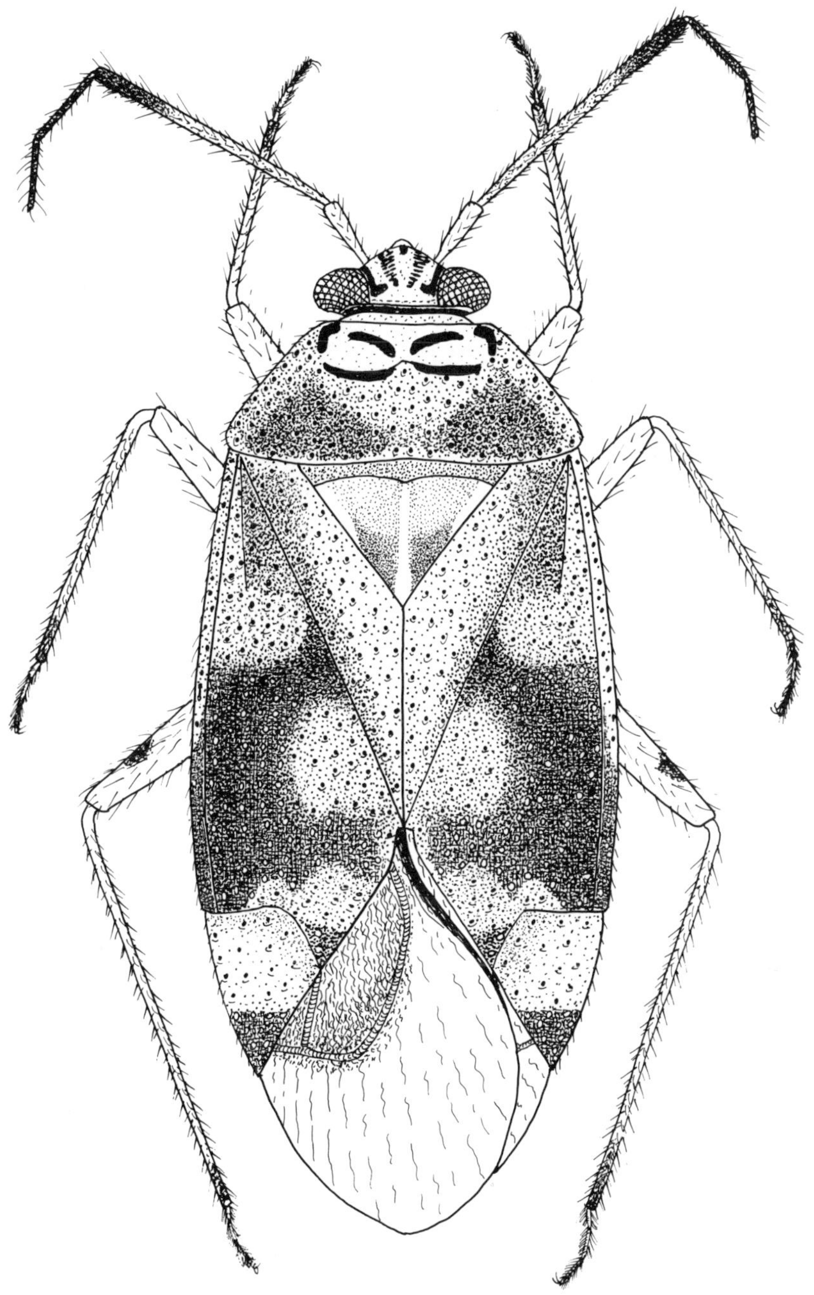

Fig. 114. *Deraeocoris nitenatus*

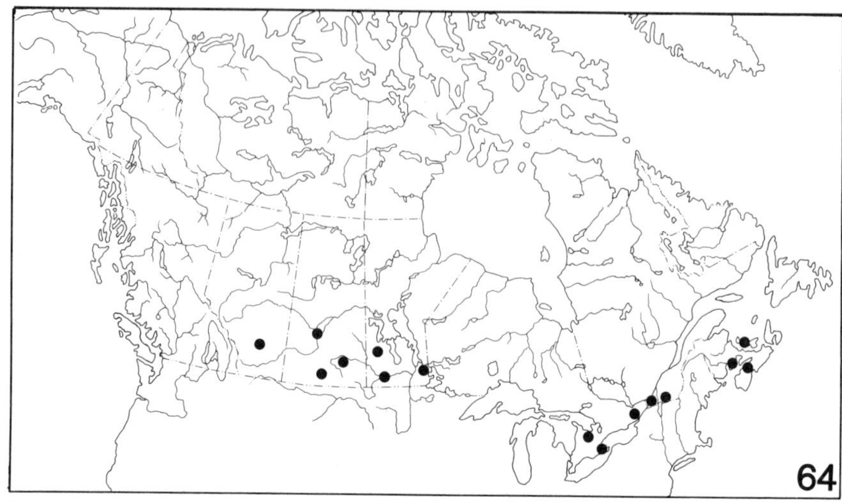

Map 64. Collection localities for *Deraeocoris aphidiphagus*.

Also collected on *Quercus* spp., and *Ulmus americana*.

Distribution.   Eastern USA; Maritime Provinces, Quebec, Ontario, Prairie Provinces (Map 64).

Deraeocoris fasciolus Knight

Figs. 116, 127; Map 65

Deraeocoris fasciolus Knight, 1921a:123.

Length 6.3–7.0 mm; width 2.9–3.2 mm. Head pale brown marked with black bars. Pronotum light brown, punctures black; calli marked with irregular black bars. Scutellum light brown, broad area each side of median line black. Hemelytra reddish brown; corium with black spot at middle; cuneus pale, tip black; wing membrane with large rounded fuscous spot at apex. Ventral surface reddish brown.

Remarks.   This species is distinguished by the reddish brown color, by the irregular black markings on the calli, by the rounded fuscous spot on the wing membrane (Fig. 116), and by the male claspers (Fig. 127).

Collected on apple in Nova Scotia and Quebec; on apple and mulberry in Ontario; on pin cherry and raspberry in the Prairie Provinces; on apple and pear in British Columbia; predaceous on aphids, psyllids, and mites. Lord (1949) observed the species preying on apple aphids, and MacPhee and Sanford (1954) observed it preying on mites, aphids, codling moth, and eyespotted bud moth eggs and larvae in Nova Scotia;

158

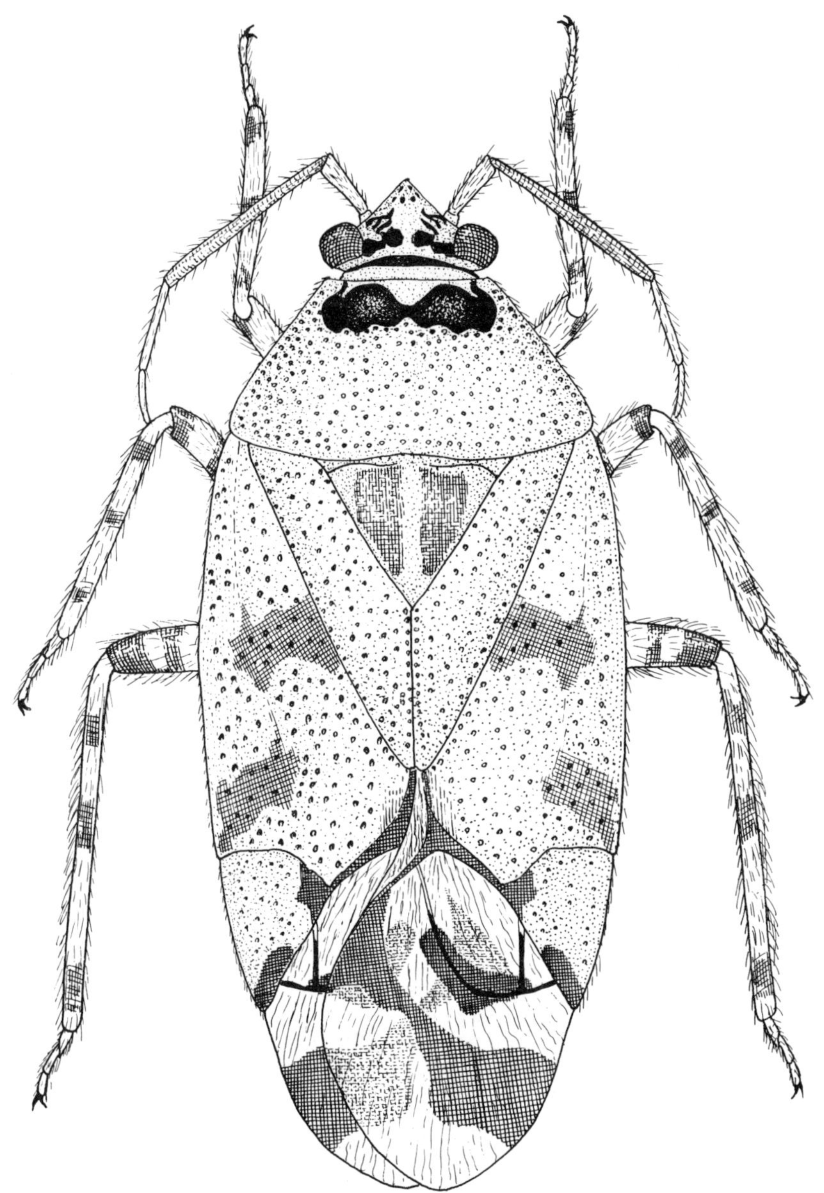

Fig. 115. *Deraeocoris aphidiphagus*

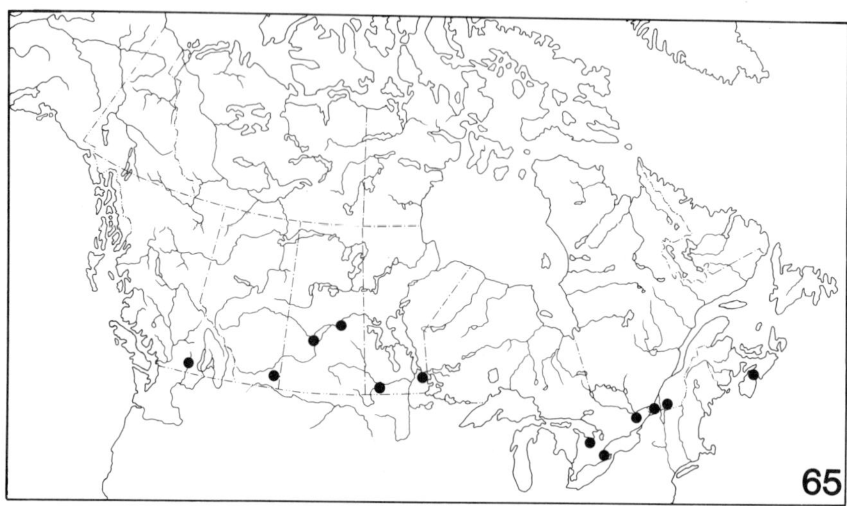

Map 65. Collection localities for *Deraeocoris fasciolus.*

Braimah et al. (1981) observed the species on apple in Quebec. McMullen and Jong (1967) observed it preying on mites and psyllids in British Columbia.

Overwinters in the egg stage. The life history is similar to that of *nitenatus*.

Also collected on *Alnus* spp., *Betula* spp., *Salix* spp., *Acer negundo*, and *Quercus macrocarpa*.

Distribution.   Northeastern and north central USA, Oregon; Nova Scotia, Quebec, Ontario, Prairie Provinces, British Columbia (Map 65).

Deraeocoris borealis (Van Duzee)

Figs. 117, 128; Map 66

Camptobrochis borealis Van Duzee, 1920:354.
Deraeocoris borealis: Knight, 1921a:120.

Length 5.7–7.0 mm; width 2.6–2.9 mm. Head light brown marked with black bars. Pronotum light brown, punctures black; calli black with black rays behind. Scutellum yellow with broad area each side of median line black. Hemelytra light brown; cuneus pale, tip black; wing membrane with rounded spot at apex as in *fasciolus*. Ventral surface reddish brown.

160

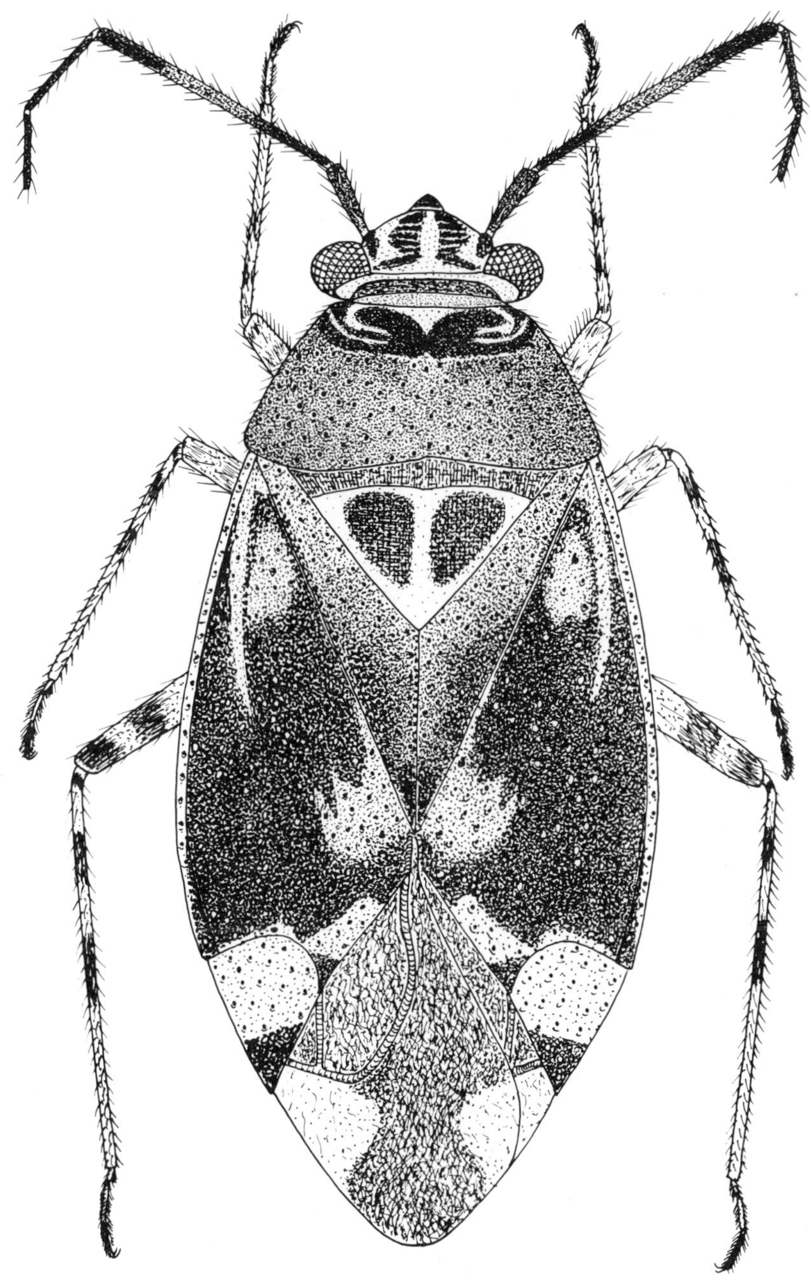

Fig. 116. *Deraeocoris fasciolus*

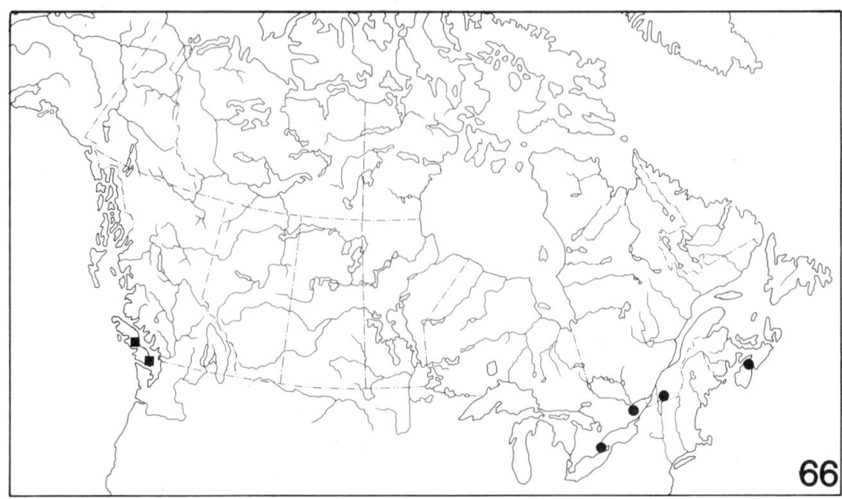

Map 66. Collection localities for *Deraeocoris borealis* (●), and *Campyloneura virgula* (■).

Remarks. This species is similar to *fasciolus* in appearance. It is separated from *fasciolus* by the completely black calli and the black rays behind them (Fig. 117), and by the differences in the right clasper (Fig. 128).

Collected on apple in Nova Scotia and Ontario; on pin cherry in Quebec; predaceous on aphids. Caesar (1920) reported the species preying on apple aphids in Ontario.

Overwinters in the egg stage. The life history is similar to that of *nitenatus*.

Also collected on *Alnus rugosa*.

Distribution. Northeastern USA; Nova Scotia, Quebec, Ontario (Map 66).

Subfamily Dicyphinae Reuter

The following are the subfamily characteristics: 1) large pronotal collar; 2) slender delicate form; 3) second segment of hind tarsus longer than first; 4) tarsal claws sharply angled at their bases; 5) hairlike parempodia; and 6) large pulvilli.

The subfamily is represented by four genera and seven species. Five species are predaceous, two species are predaceous and phytophagous.

162

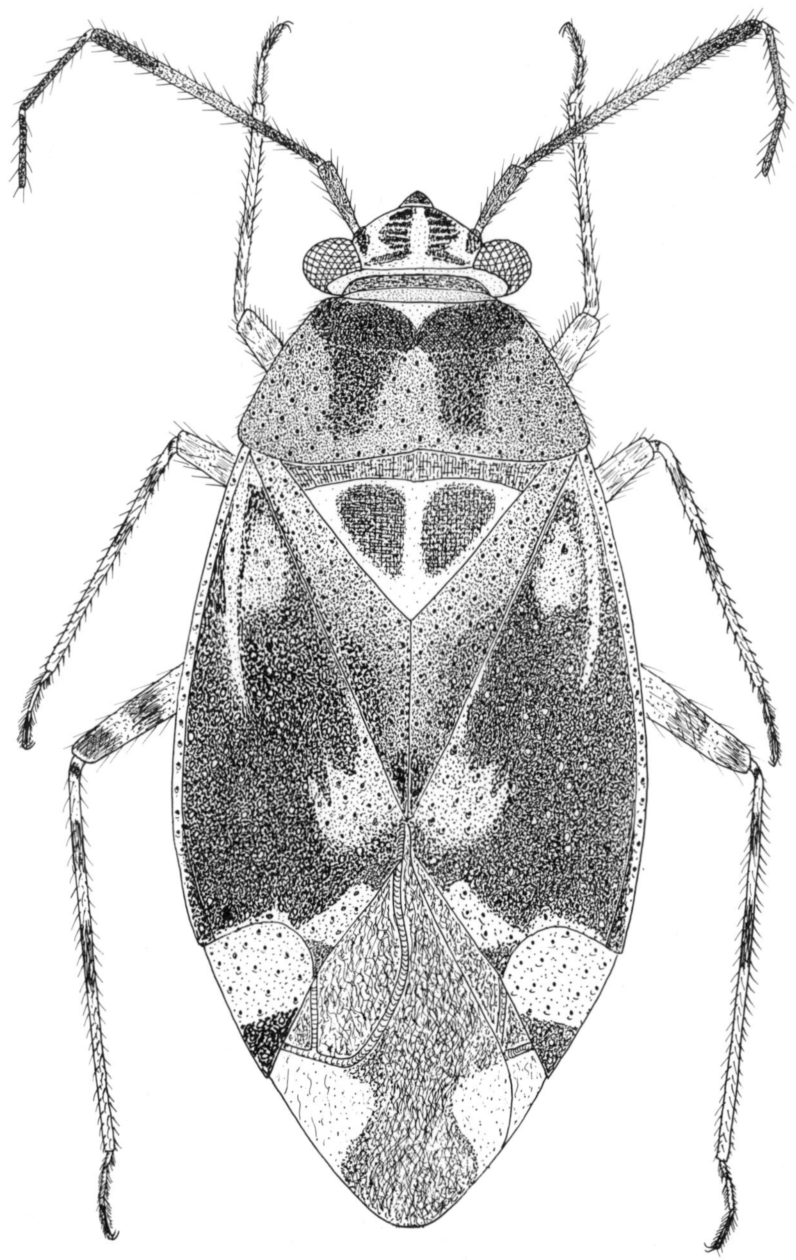

Fig. 117. *Deraeocoris borealis*

Key to genera of Dicyphinae

1. Eyes close to pronotum (Figs. 118,119) 2
   Eyes remote from pronotum (Fig. 120) 3
2. Second antennal segment much longer than width of pronotum (Fig. 118)
   .... ***Campyloneura* Fieber** (p. 164)
   Second antennal segment shorter than width of pronotum (Fig. 119) .....
   .... ***Cyrtopeltis* Fieber** (p. 165)
3. Eyes small (Fig. 120) .... ***Macrolophus* Fieber** (p. 168)
   Eyes large (Figs. 121–124) .... ***Dicyphus* Fieber** (p. 170)

Genus *Campyloneura* Fieber

Elongate, slender species. Head vertical; eyes large, close to pronotum, carina between them absent. Pronotum, scutellum, and hemelytra impunctate; pubescence simple.

One species, introduced from Europe, was collected. Overwinters in the egg stage.

Campyloneura virgula (Herrich-Schaeffer)

Fig. 118; Map 66

Capsus virgula Herrich-Schaeffer, 1836:51.
Campyloneura virgula: Fieber, 1861:268.

Length 4.0–4.7 mm; width 1.2–1.4 mm. Head black, spot behind each eye yellow. First antennal segment mostly pale yellow, second segment black, longer than width of pronotum at base. Pronotum pale green, calli orange or reddish. Scutellum yellow. Hemelytra pale translucent; clavus often brown, cuneus yellow, apex red. Ventral surface pale yellow, side of abdomen with black spot near apex; legs pale yellow.

Remarks. Downes (1957) first reported this European species from British Columbia. It is distinguished by the black head, by the orange or red calli, and by the yellow cuneus with a red apex (Fig. 118).

Collected on plum in British Columbia; predaceous on mites. Only the females were collected.

The nymphs appear in May and the adults in early June. The females are active from June to September, gradually dying out by the end of September.

Also collected on *Acer macrophyllum*, *Corylus californica*, *Ligustrum vulgare*, and *Rosa nutkana*.

Distribution. Europe; British Columbia (Map 66).

Genus *Cyrtopeltis* Fieber

Elongate, slender species. Head oblique, eyes large, close to pronotum, carina between them absent. Pronotum finely rugose. Hemelytra impunctate; pubescence simple.

One species was collected. Adults hibernate.

Cyrtopeltis bakeri Knight

Figs. 119, 129; Map 67

Cyrtopeltis bakeri Knight, 1943:58.

Length 3.5–3.7 mm; width 1.0–1.1 mm. Head black, spot near each eye on vertex yellow. First antennal segment black, base and apex yellow; second segment yellow, base and apex fuscous, shorter than width of pronotum at base. Pronotum and scutellum black. Hemelytra black; costal margin pale green, cuneus pale green, spot near apex black. Ventral surface black; legs pale green.

Remarks. This species is distinguished by the black head, pronotum, and scutellum, by the black hemelytra with pale green costal margins (Fig. 119), and by the sickle-shaped left clasper (Fig. 129).

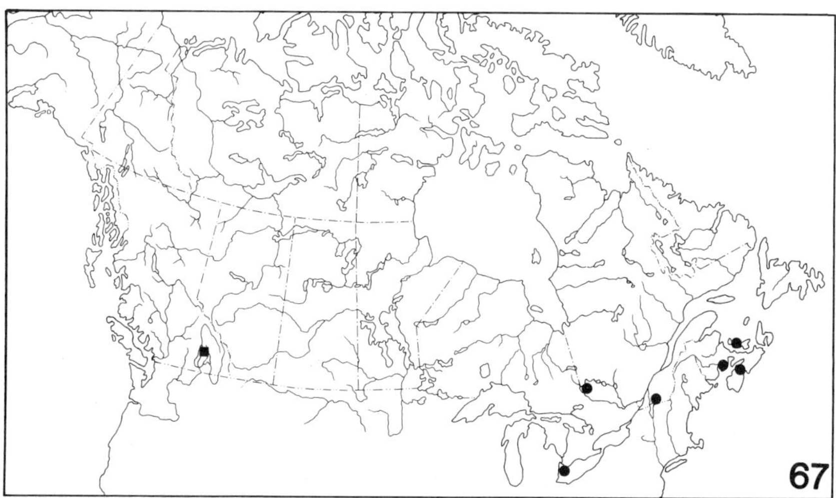

Map 67. Collection localities for *Cyrtopeltis bakeri* (■), and *Macrolophus tenuicornis* (●).

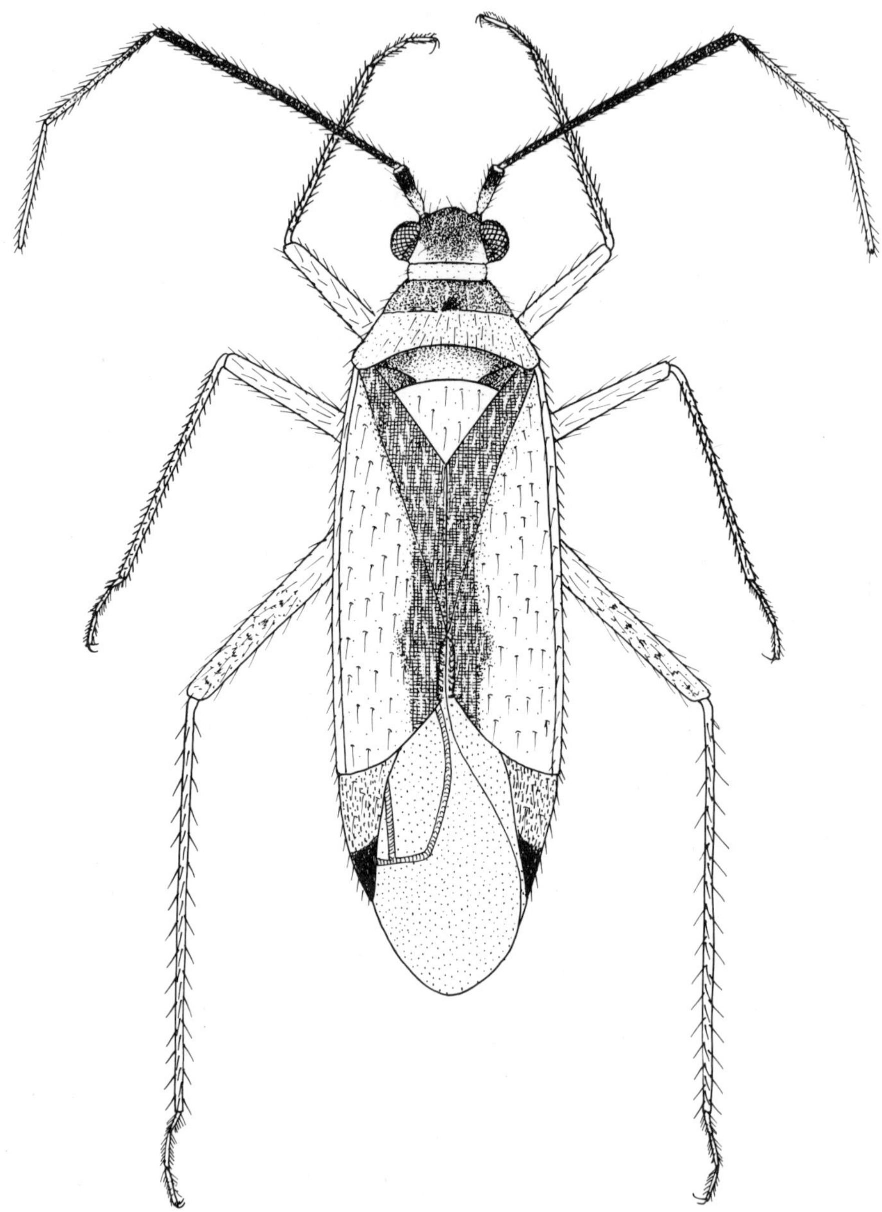

Fig. 118. *Campyloneura virgula*

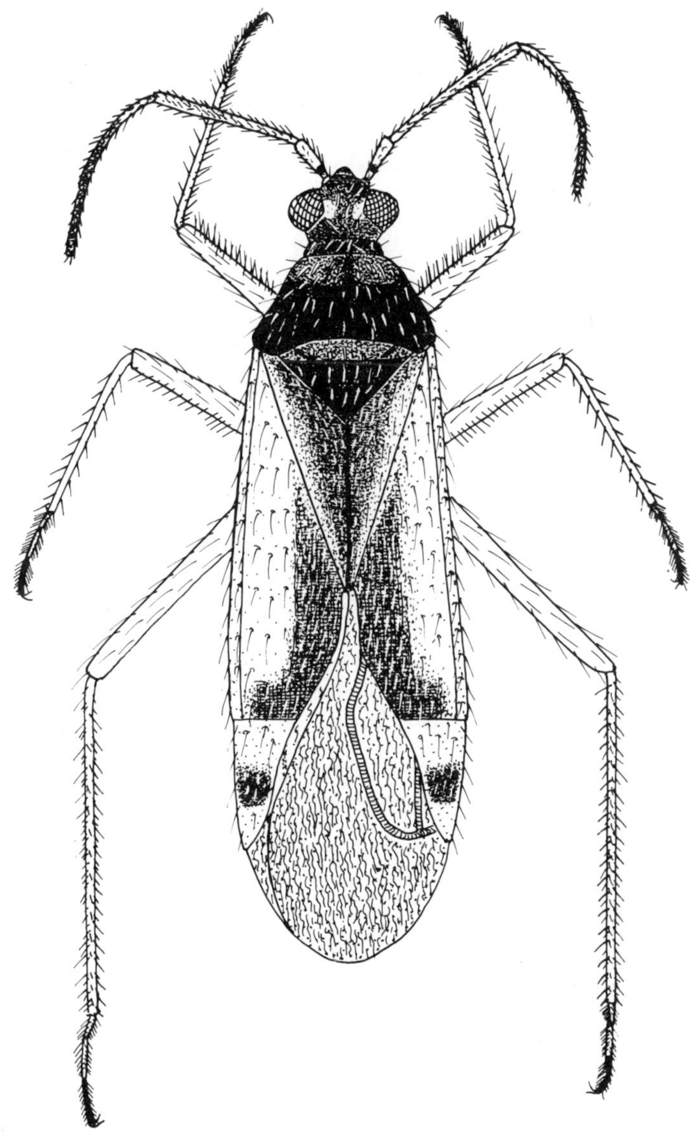

Fig. 119. *Cyrtopeltis bakeri*

Collected on thimbleberry in British Columbia; predaceous on aphids.

The overwintered adults lay eggs early in the spring, and gradually die out by the end of June. The nymphs appear about the end of May and the new adults about the end of June. The new generation adults are active until hibernation.

Distribution.   Oregon, Washington; British Columbia (Map 67).

Genus *Macrolophus* Fieber

Elongate, slender species. Head horizontal, front declivent; eyes small, far removed from pronotum; carina between eyes absent. Pronotum and hemelytra impunctate; pubescence simple. Legs long and slender.

One species was collected. Overwinters in the egg stage.

Macrolophus tenuicornis Blatchley

Fig. 120; Map 67

Macrolophus tenuicornis Blatchley, 1926:913.
Macrolophus longicornis Knight, 1926:314.

Length 4.0–4.4 mm; width 0.9–1.1 mm. Head green, stripe behind each eye black; first antennal segment mostly black, longer than width of head; second segment green, apex black, longer than width of pronotum at base. Pronotum and scutellum green. Hemelytra green; clavus and inner corium spotted with fuscous, outer corium with large black spot near apex; cuneus pale green, apex black. Ventral surface and legs green.

Remarks.   This species is distinguished by the small eyes, by the long first antennal segments, and by the spotting on the hemelytra (Fig. 118).

Collected on raspberry in the Maritime Provinces, Quebec, and Ontario; predaceous on aphids.

The nymphs appear about mid-May and the adults about mid-June. The adults are active throughout July and August, and gradually die out by early September.

Also collected on *Dennstaedtia punctilobula*, *Polymnia canadensis*, *Aster* spp., *Gerardia pectinata*, and *Geranium* sp.

Distribution.   Eastern USA; Maritime Provinces, Quebec, Ontario (Map 67).

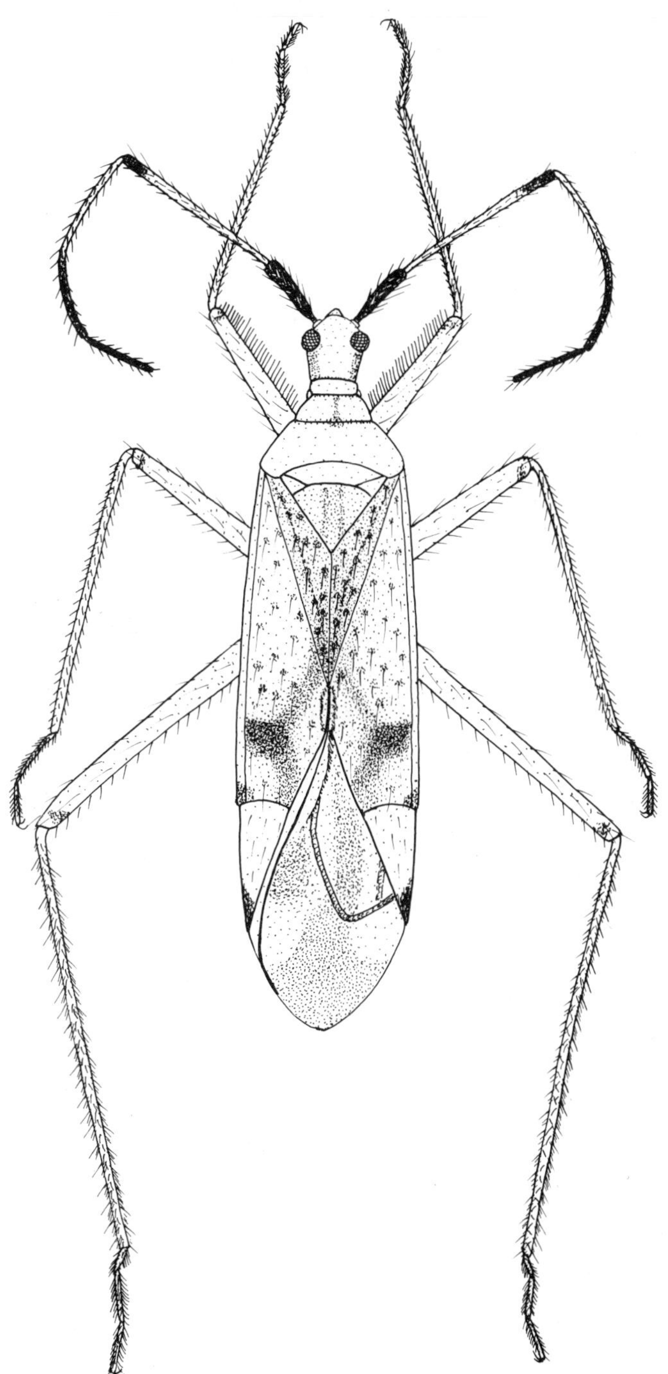

Fig. 120. *Macrolophus tenuicornis*

Genus *Dicyphus* Fieber

Elongate, slender species. Head nearly vertical in front; eyes large, considerably removed from pronotum; carina between eyes absent. Pronotum impunctate, calli prominent. Hemelytra impunctate; pubescence simple. Legs long, slender.

Four species were collected; three species hibernate, one overwinters in the egg stage.

Key to species of *Dicyphus*

1. Osteolar peritreme depressed, inconspicuous; pronotum black or with yellow median line (Fig. 121); left clasper (Fig. 130) ..... *rubi* **Knight** (p. 170)
   Osteolar peritreme elevated, prominent; pronotum mostly yellowish or reddish 2
2. Larger species, over 4.4 mm long (Fig. 122); left clasper (Fig. 131) .... *famelicus* **(Uhler)** (p. 172)
   Small species, under 4.0 mm long 3
3. Second antennal segment longer than width of pronotum at base (Fig. 123); left clasper (Fig. 132) .... *discrepans* **Knight** (p. 174)
   Second antennal segment shorter than width of pronotum at base (Fig. 124); left clasper (Fig. 133) .... *hesperus* **Knight** (p. 176)

Dicyphus rubi Knight

Figs. 121, 130; Map 68

Dicyphus rubi Knight, 1968:72.

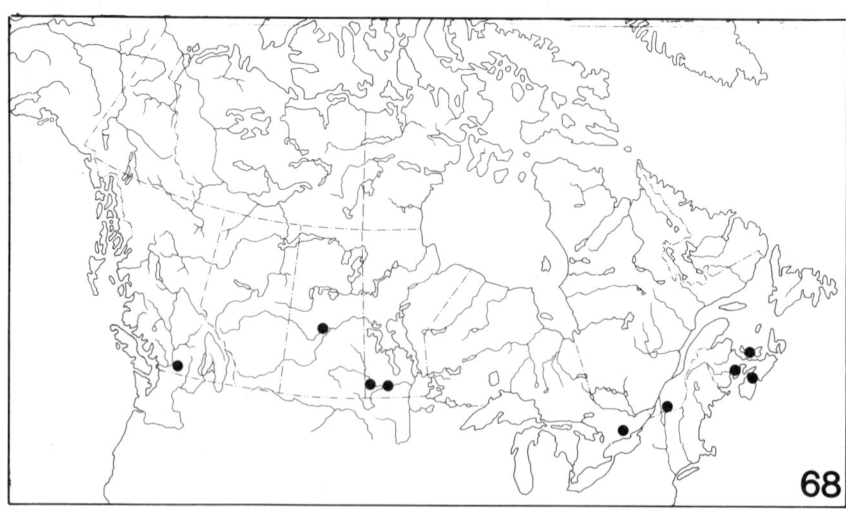

Map 68. Collection localities for *Dicyphus rubi*.

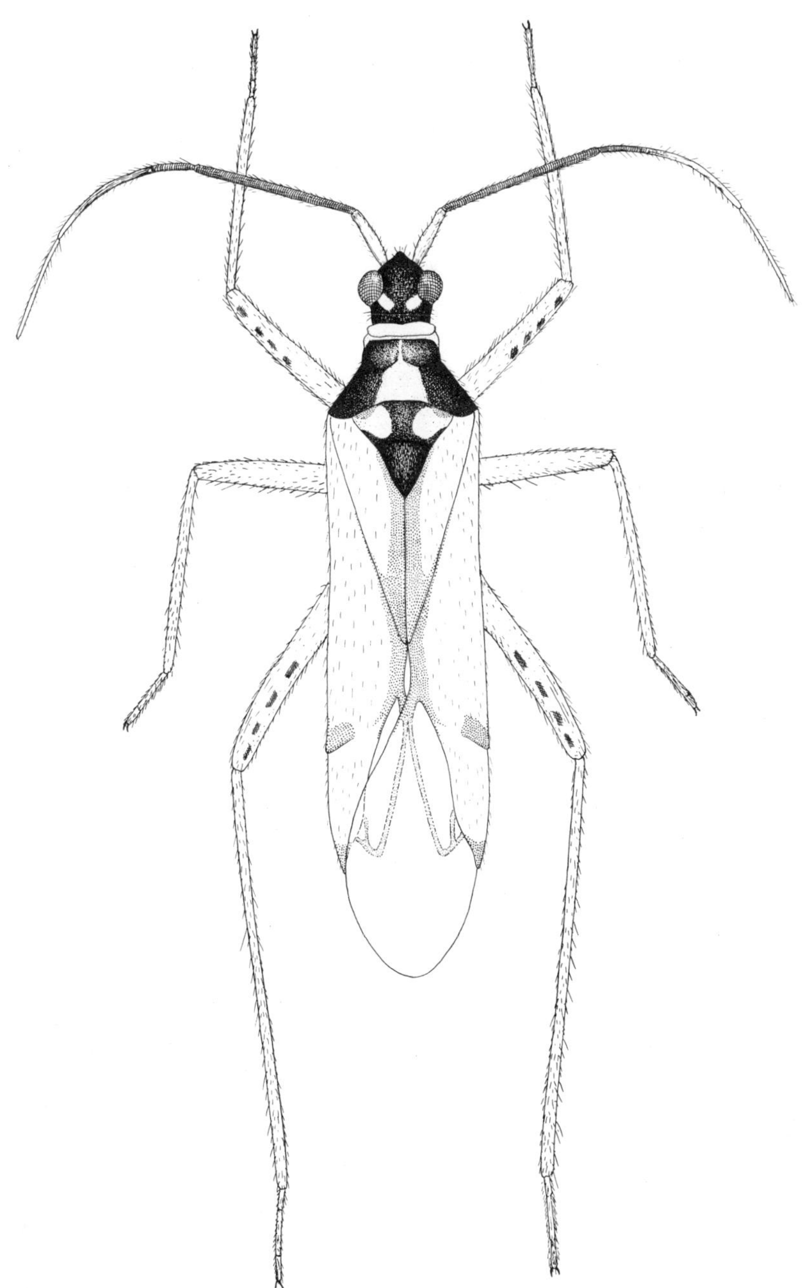

Fig. 121. *Dicyphus rubi*

Length 4.2–4.6 mm; width 0.9–1.1 mm. Head black, spot on top next to each eye yellow. First antennal segment pale, second segment black. Pronotum black, wedge-shaped median area often green; collar white. Scutellum black. Hemelytra pale green; inner clavus fuscous, spot at apex of embolium and at apex of cuneus fuscous. Ventral surface black; legs green.

Remarks.   This species is distinguished by the black scutellum, by the pale first antennal segment, by the black second antennal segment (Fig. 121), and by the slender process on the left clasper (Fig. 130).

Collected on raspberry in the Maritime Provinces, Quebec, Ontario, and Prairie Provinces; on squashberry, thimbleberry, raspberry, and gooseberry in British Columbia; phytophagous and predaceous on aphids.

Overwinters in the egg stage. The nymphs appear in early May and the adults in early June. The adults are active throughout July and August, and gradually die out by mid-September.

Distribution.   New York, Michigan, Colorado, Utah; Maritime Provinces, Quebec, Ontario, Prairie Provinces, British Columbia (Map 68).

Dicyphus famelicus (Uhler)

Figs. 122, 131; Map 69

Idolocoris famelicus Uhler, 1878:413.
Dicyphus famelicus: Atkinson, 1890:128.

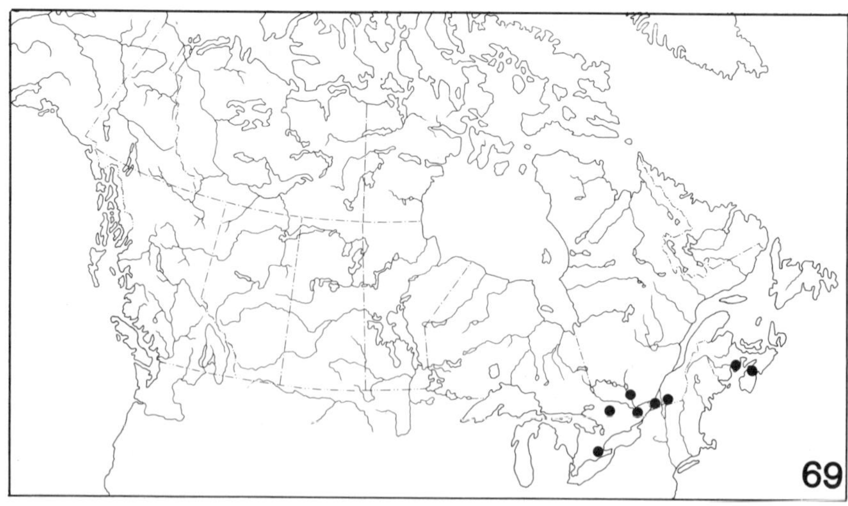

Map 69. Collection localities for *Dicyphus famelicus*.

Length 4.4–4.7 mm; width 1.0–1.2 mm. Head light yellow often marked with red. First antennal segment pale green, bar near base and apex red; second segment pale green, apical one third red or black. Pronotum and scutellum light yellow. Hemelytra opaque white marked with orange or red. Ventral surface yellow or orange; legs pale green. Macropterous.

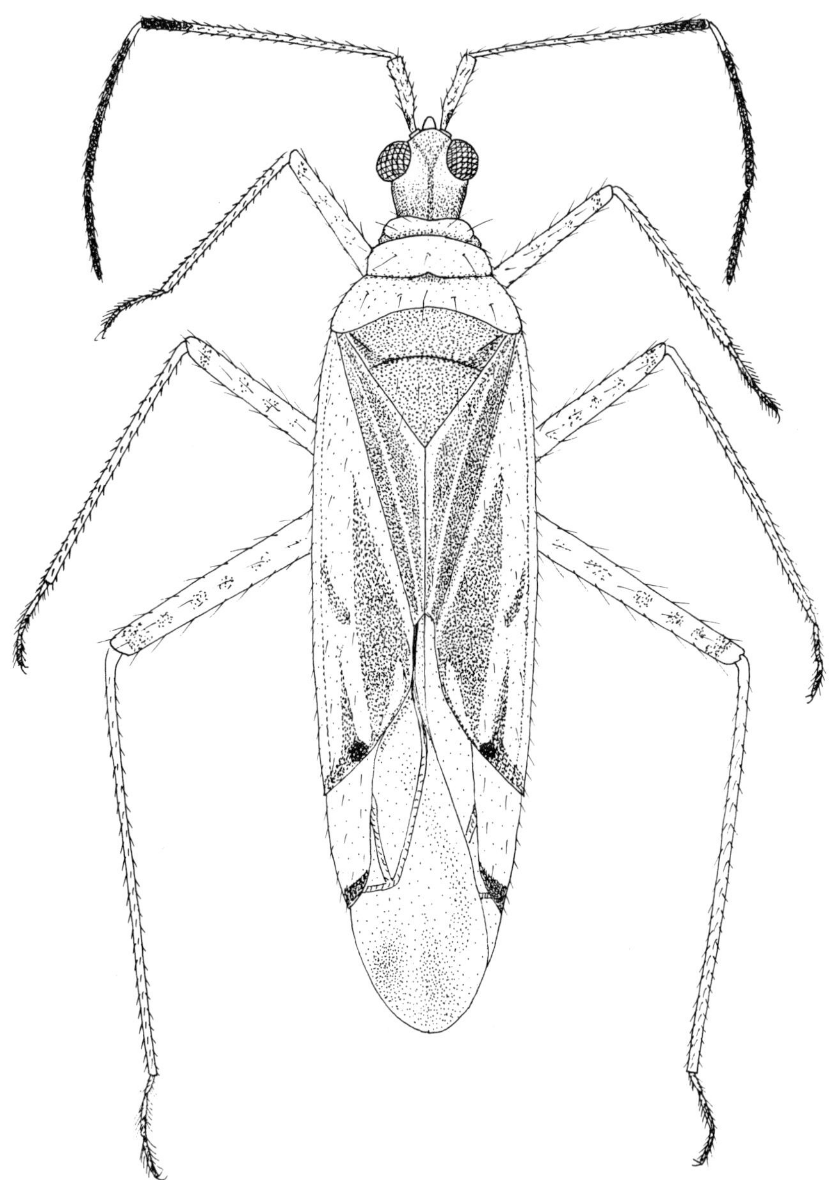

Fig. 122. *Dicyphus famelicus*

Remarks. This species is distinguished by the large size, by the reddish or orange color on the hemelytra (Fig. 122), and by the left clasper (Fig. 131).

Collected on raspberry in Nova Scotia, New Brunswick, Quebec, and Ontario; phytophagous, and predaceous on aphids.

The adults hibernate. The eggs are laid in the spring, and the adults gradually die out by the end of June. The nymphs appear about the first part of June and the new adults about the first part of July. The new adults are active until hibernation.

Distribution. Eastern USA; Nova Scotia, New Brunswick, Quebec, Ontario (Map 69).

Dicyphus discrepans Knight

Figs. 123, 132; Map 70

Dicyphus discrepans Knight, 1923*b*:476.

Length 3.2–3.9 mm; width 0.9–1.1 mm. Head black, area behind each eye yellow. Rostrum 1.4–1.6 mm long. First antennal segment yellow, base and apex red, often all black; second segment longer than width of pronotum at base. Pronotum pale yellow, transverse groove behind calli often red. Scutellum yellow, median longitudinal line reddish brown.

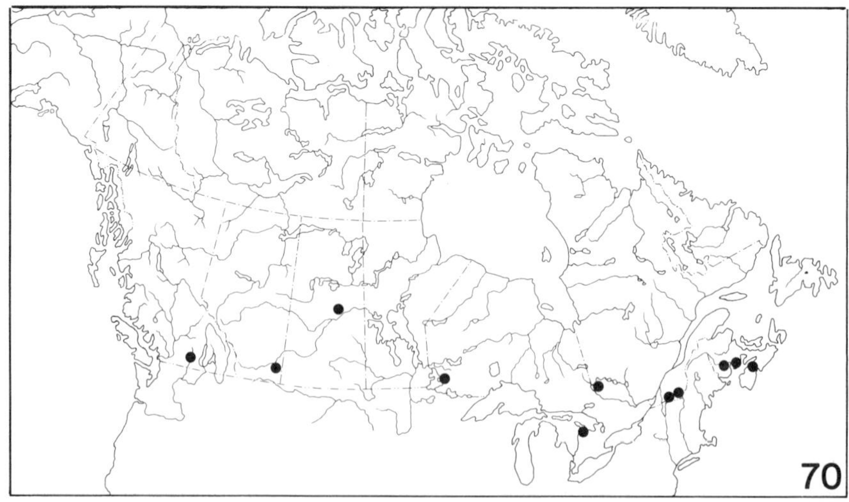

Map 70. Collection localities for *Dicyphus discrepans*.

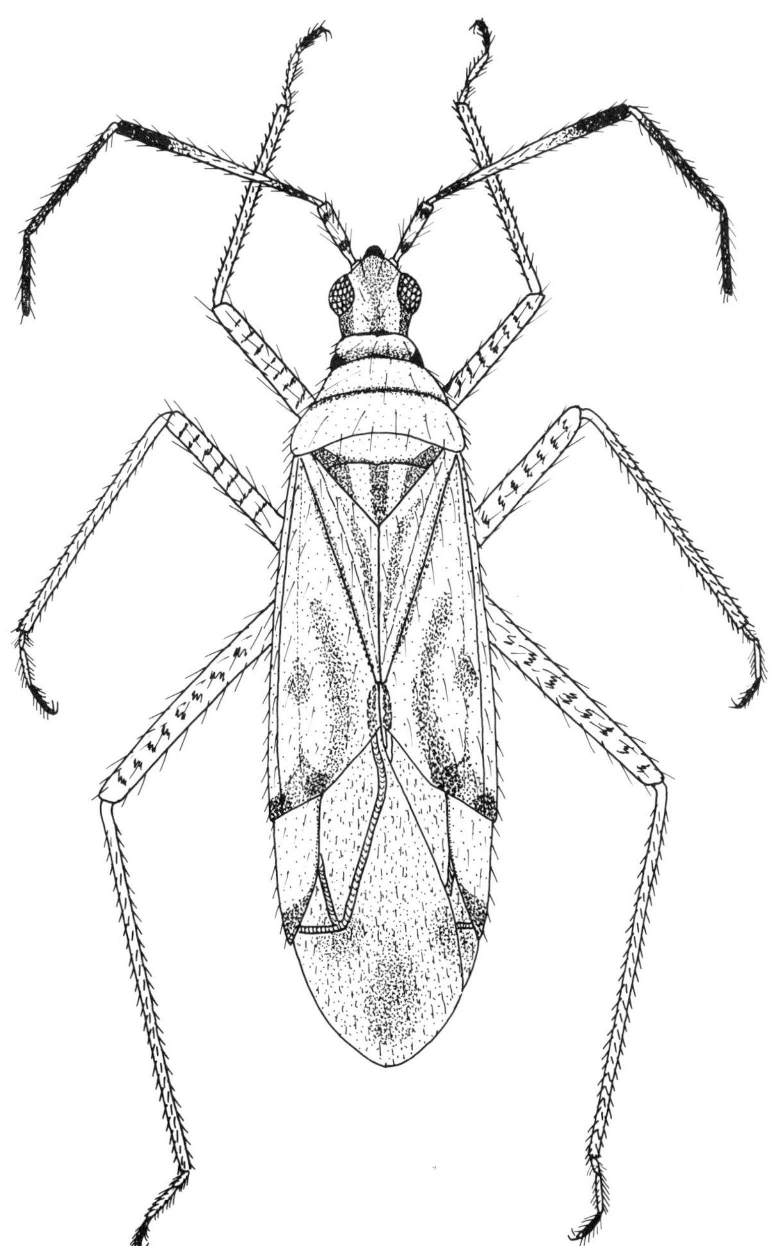

Fig. 123. *Dicyphus discrepans*

Hemelytra opaque white marked with fuscous; apex of corium with black spot, apex of cuneus red. Ventral surface black; legs pale green. Brachypterous and macropterous.

Remarks. This species is distinguished by the black head (Fig. 123) and the black ventral surface, and by the left clasper (Fig. 132).

Collected on raspberry in Nova Scotia, New Brunswick, Quebec, Ontario, and Prairie Provinces; on thimbleberry and gooseberry in British Columbia; predaceous on aphids.

The adults hibernate. The life history is similar to that of *famelicus*.

Also collected on *Aster* spp., *Salix* spp., and *Rosa* spp.

Distribution. Northern USA; Nova Scotia, New Brunswick, Quebec, Ontario, Prairie Provinces, British Columbia (Map 70).

Dicyphus hesperus Knight

Figs. 124, 133; Map 71

Dicyphus hesperus Knight, 1943:56.

Length 3.2–3.9 mm; width 0.9–1.1 mm. Rostrum 1.2–1.4 mm long. First antennal segment black, middle area often yellow; second segment shorter than width of pronotum at base. Similar to *discrepans* in size and color. Males macropterous, females macropterous and brachypterous.

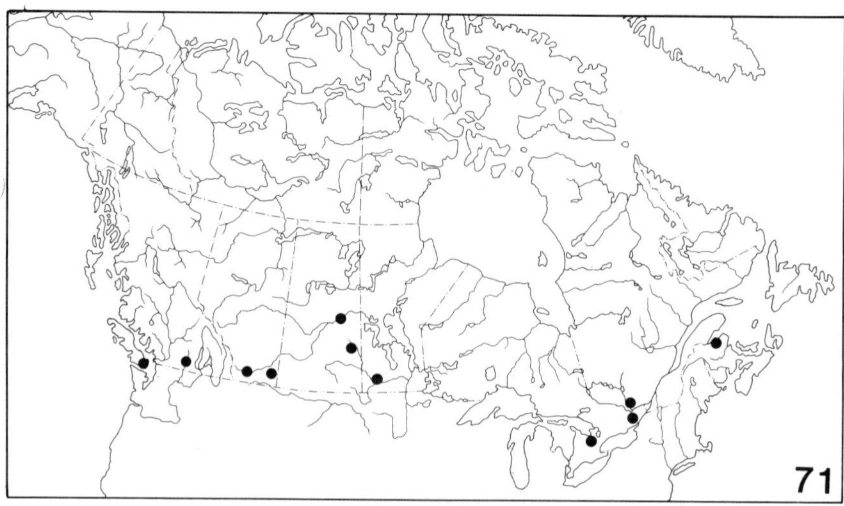

Map 71. Collection localities for *Dicyphus hesperus*.

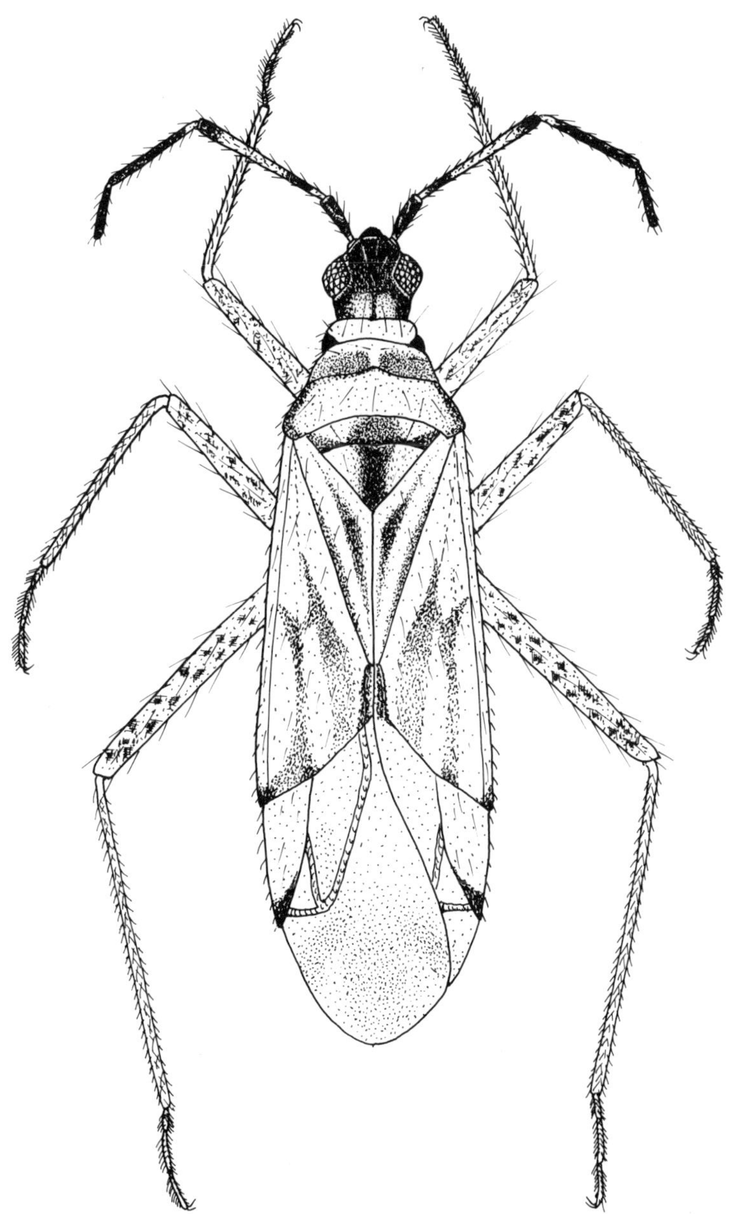

Fig. 124. *Dicyphus hesperus*

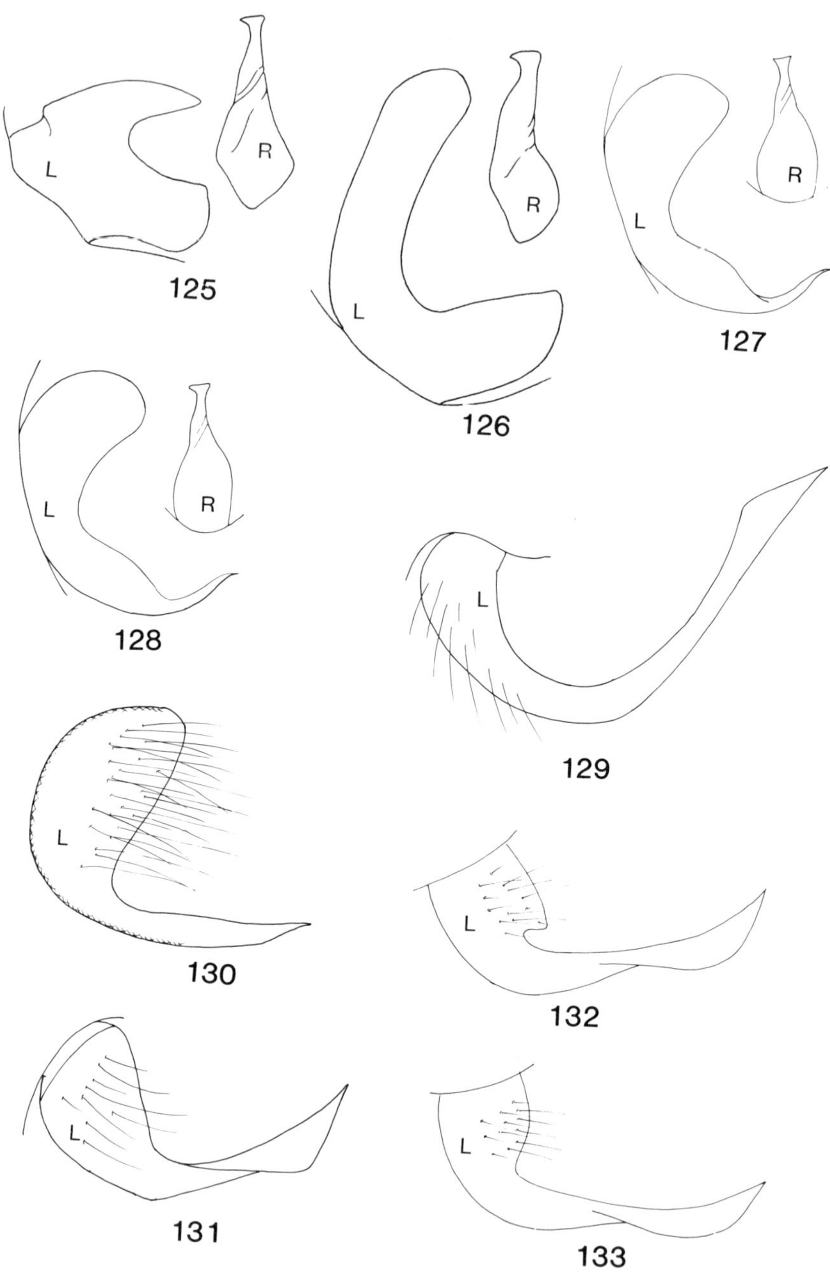

Figs. 125–133. Male claspers of Deraeocorinae and Dicyphinae. 125, *Deraeocoris nitenatus*; 126, *Deraeocoris aphidiphagus*; 127, *Deraeocoris fasciolus*; 128, *Deraeocoris borealis*; 129, *Cyrtopeltis bakeri*; 130, *Dicyphus rubi*; 131, *Dicyphus famelicus*; 132, *Dicyphus discrepans*; 133, *Dicyphus hesperus*.

178

Remarks. This species is separated from *discrepans* by the shorter second antennal segment (Fig. 124) and by the absence of the notch on the left clasper (Fig. 133).

Collected on raspberry, blackberry, and thimbleberry in British Columbia; on raspberry in the Prairie Provinces; on raspberry and thimbleberry in Ontario, Quebec, and New Brunswick; predaceous on aphids.

The adults hibernate. The life history is similar to that of *famelicus*.

Also collected on *Mentha arvensis*.

Distribution. Western USA; British Columbia, Prairie Provinces, Ontario, Quebec, New Brunswick (Map 71).

Scientific and common names of plants

Abies balsamea	balsam fir
Abies lasiocarpa	alpine fir
Acer macrophyllum	broadleaf maple
Acer negundo	Manitoba maple
Acer saccharum	sugar maple
alfalfa	*Medicago sativa*
Allegheny serviceberry	*Amelanchier laevis*
alpine fir	*Abies lasiocarpa*
Alnus rugosa	speckled alder
Ambrosia trifida	giant ragweed
Amelanchier alnifolia	saskatoon
Amelanchier laevis	Allegheny serviceberry
American beech	*Fagus grandifolia*
American elm	*Ulmus americana*
American hazelnut	*Corylus americana*
apple	*Malus* spp.
apricot	*Prunus armeniaca*
American elm	*Ulmus americana*
Artemisia tridentata	big sagebrush
balsam fir	*Abies balsamea*
basswood	*Tilia americana*
Betula occidentalis	water birch
Betula papyrifera	paper birch
big sagebrush	*Artemisia tridentata*
blackberry	*Rubus ursinus*
black ash	*Fraxinus nigra*
black cherry	*Prunus serotina*
black locust	*Robinea pseudacacia*
black walnut	*Juglans nigra*
blue beech	*Carpinus caroliniana*
blueberry	*Vaccinium* spp.
bog-laurel	*Kalmia polifolia*
broom	*Cytisus scoparius*
broadleaf maple	*Acer macrophyllum*
bur oak	*Quercus macrocarpa*
California hazelnut	*Corylus californica*
caragana	*Caragana arborescens*
Caragana arborescens	caragana
Carpinus caroliniana	blue beech
Carya ovata	shagbark hickory
Carya spp.	hickory
Ceanothus sanguineus	redstem ceanothus
chokecherry	*Prunus virginiana*
chrysanthemum	*Chrysanthemum* spp.

Chrysanthemum spp. — chrysanthemum
Chrysothamnus nauseosus — rabbitbrush
Corylus americana — American hazelnut
Corylus avellana — European hazel
Corylus californica — California hazelnut
cranberry — *Vaccinium macrocarpon*
Crataegus chrysocarpa — fireberry
common mullein — *Verbascum thapsus*
common privet — *Ligustrum vulgare*
common wild rose — *Rosa nutkana*
common witch-hazel — *Hamamelis virginiana*
currant — *Ribes* spp.
Cytisus scoparius — broom

Dennstaedtia punctilobula — hayscented fern

elderberry — *Sambucus glauca*
European ash — *Fraxinus excelsior*
European beech — *Fagus sylvatica*
European hazel — *Corylus avellana*

Fagus grandifolia — American beech
Fagus sylvatica — European beech
field mint — *Mentha arvensis*
fireberry — *Crataegus chrysocarpa*
Fragaria spp. — strawberry
Fraxinus americana — white ash
Fraxinus excelsior — European ash
Fraxinus nigra — black ash
Fraxinus pennsylvanica — red ash

geranium — *Geranium* spp.
Geranium spp. — geranium
gerardia — *Gerardia pectinata*
Gerardia pectinata — gerardia
giant ragweed — *Ambrosia trifida*
gooseberry — *Ribes* spp.
grape, cultivated — *Vitis* spp.
grape, wild — *Vitis rotundifolia*
greasewood — *Purshia tridentata*

Hamamelis virginiana — common witch-hazel
hayscented fern — *Dennstaedtia punctilobula*
hickory — *Carya* spp.
high bush-cranberry — *Viburnum trilobum*

ironwood — *Ostrya virginiana*

Juglans nigra — black walnut

Kalmia polifolia — bog-laurel

leafcup — *Polymnia canadensis*
Ligustrum vulgare — common privet
lodgepole pine — *Pinus contorta*
loganberry — *Rubus* spp.

Malus spp. — apple
Manitoba maple — *Acer negundo*
Medicago sativa — alfalfa
Mentha arvensis — field mint
Morus rubra — red mulberry

nannyberry — *Viburnum lentago*

oak — *Quercus* spp.
Ostrya virginiana — ironwood

paper birch — *Betula papyrifera*
Picea glauca — white spruce
peach — *Prunus persica*
pear — *Pyrus communis*
pin cherry — *Prunus pensylvanica*
Pinus contorta — lodgepole pine
plum — *Prunus domestica*
Polymnia canadensis — leafcup
potato — *Solanum* spp.
prickly ash — *Zanthoxylum americanum*
Prunus armeniaca — apricot
Prunus avium — sweet cherry
Prunus cerasus — sour cherry
Prunus domestica — plum
Prunus pensylvanica — pin cherry
Prunus persica — peach
Prunus serotina — black cherry
Prunus virginiana — chokecherry
Purshia tridentata — greasewood
pyramidal spirea — *Spiraea pyramidea*

Quercus alba — white oak
Quercus macrocarpa — bur oak
Quercus rubra — red oak

rabbitbrush — *Chrysothamnus nauseosus*
raspberry — *Rubus* spp.
red ash — *Fraxinus pennsylvanica*
red mulberry — *Morus rubra*
red oak — *Quercus rubra*
redstem ceanothus — *Ceanothus sanguineus*
Ribes spp. — currant

Ribes spp.	gooseberry
Robinea pseudacacia	black locust
Rosa nutkana	common wild rose
Rubus spp.	loganberry
Rubus spp.	raspberry
Rubus parviflorus	thimbleberry
Rubus ursinus	blackberry
russet buffaloberry	*Shepherdia canadensis*
Salix interior	sandbar willow
Sambucus glauca	elderberry
sandbar willow	*Salix interior*
saskatoon	*Amelanchier alnifolia*
shagbark hickory	*Carya ovata*
Shepherdia canadensis	russet buffaloberry
small-leaved lime	*Tilia cordata*
Solanum spp.	potato
sour cherry	*Prunus cerasus*
speckled alder	*Alnus rugosa*
Spiraea pyramidea	pyramidal spirea
squashberry	*Viburnum edule*
strawberry	*Fragaria* spp.
sugar maple	*Acer saccharum*
sweet cherry	*Prunus avium*
thimbleberry	*Rubus parviflorus*
Tilia americana	basswood
Tilia cordata	small-leaved lime
Ulmus americana	American elm
Vaccinium spp.	blueberry
Vaccinium macrocarpon	cranberry
Verbascum thapsus	common mullein
Viburnum edule	squashberry
Viburnum lentago	nannyberry
Viburnum trilobum	high bush-cranberry
Vitis spp.	grape, cultivated
Vitis rotundifolia	grape, wild
water birch	*Betula occidentalis*
white ash	*Fraxinus americana*
white oak	*Quercus alba*
white spruce	*Picea glauca*
Zanthoxylum americanum	prickly ash

Glossary

Structures labeled in Figure 1 are not repeated here.

apex   That part of any joint or structure opposite the base by which it is attached.
appressed   Pressed close to or lying flat.
apterous   Lacking wings.
brachypterous   With short or abbreviated wings.
claspers   The left and right structures on the male genital segment used in copulation.
clavate   Club-shaped.
convergent   Coming together.
convex   Rounded outward like the outside of a bowl.
costal   The lateral margin of the hemelytron.
declivent   Sloping gradually downward.
divergent   Drawing apart.
dorsum   Dorsal surface.
ductus seminis   The seminal duct of the male genitalia.
dusky   Somewhat dark in color.
fuscous   Dusky; brownish gray, approaching black.
glabrous   Hairs present, but too short to be seen readily.
horizontal   Relating to the horizontal plane.
impunctate   Without punctures.
incrasssate   Thickened toward the apex.
irrorate   Speckled.
macropterous   Wings fully developed.
nymph   An immature mirid.
oblique   Inclined, sloping.
pilose   Covered with fine, long hairs.
pleuron   The side of the thorax.
predaceous   Preying on other insects.
pruinose   Covered with whitish dust.
pubescence   Fine, soft hair covering the surface.
punctate   Marked with small depressions.
rugose   Wrinkled.
sclerite   A hard sclerotized piece of integument, or covering.
sericeous   Silky.
spiculum   A slender sclerite.
striate   Grooved.
submedian   Below the median; e.g., a line on the frons between the median line and the eye.
trunctate   Squared rather than rounded or pointed at the tip.
tubercle   A small or moderate projection.
tumid   Swollen.
vesica   The male phallus, or intromittent organ.

Fruit crop and plant bug association

(* phytophagous; # predaceous; *# phytophagous and predaceous)

Amelanchier spp.

Allegheny serviceberry

*Lygocoris communis
*Lygocoris omnivagus
*Lygocoris quercalbae
*Lygus lineolaris

saskatoon

#Deraeocoris aphidiphagus
#Deraeocoris brevis
#Deraeocoris nitenatus
 *Lygocoris communis
 *Lygocoris quercalbae
 *Lygus lineolaris
 *Plagiognathus obscurus

Fragaria spp. — strawberry

 *Calocoris norvegicus
 *Lopidea dakota
 *Lygus lineolaris
 *Lygus varius
*#Plagiognathus chrysanthemi

Malus spp. — apple

*#Atractotomus mali
 #Blepharidopterus angulatus
*#Campylomma verbasci
  *Capsus ater
 #Ceratocapsus digitulus
 #Ceratocapsus fuscinus
 #Ceratocapsus modestus
 #Ceratocapsus pilosulus
 #Ceratocapsus pumilus
 #Deraeocoris aphidiphagus
 #Deraeocoris bakeri
 #Deraeocoris borealis
 #Deraeocoris brevis
 #Deraeocoris fasciolus
 #Deraeocoris nebulosus
 #Deraeocoris nitenatus
 #Diaphnocoris provancheri

#Eurychilopterella luridula
 *Heterocordylus malinus
#Heterotoma meriopterum
#Hyaliodes harti
#Hyaliodes vitripennis
#Lepidopsallus minisculus
 *Lygidea mendax
 *Lygocoris communis
 *Lygocoris omnivagus
 *Lygus hesperus
 *Lygus lineolaris
 *Neurocolpus nubilus
#Orthotylus viridinervis
#Paraproba capitata
#Phytocoris canadensis
#Phytocoris conspurcatus
#Phytocoris corticevivens
#Phytocoris cortitectus
#Phytocoris dimidiatus
#Phytocoris erectus
#Phytocoris gracillatus
#Phytocoris husseyi
#Phytocoris interspersus
#Phytocoris lasiomerus
#Phytocoris neglectus
#Phytocoris nigricollis
#Phytocoris onustus
#Phytocoris salicis
#Pilophorus perplexus
*#Plagiognathus obscurus
*#Plagiognathus politus
 #Psallus salicellus
 *Stenotus binotatus
 *Taedia pallidula

Morus rubra — red mulberry

#Ceratocapsus modestus
#Ceratocapsus pilosulus
#Deraeocoris fasciolus
#Deracocoris nebulosus
#Deraeocoris nitenatus
#Diaphnocoris provancheri
#Hyaliodes harti
 *Lygocoris communis
 *Lygocoris omnivagus
 *Lygus lineolaris
#Paraproba capitata

186

#Phytocoris conspurcatus
#Phytocoris onustus
*#*Plagiognathus politus*

Prunus spp.

apricot

#Ceratocapsus fuscinus
#Ceratocapsus incisus
#Ceratocapsus pilosulus
#Diaphnocoris provancheri
#Hyaliodes harti
Lygocoris caryae
Lygocoris communis
Lygocoris omnivagus
Lygus lineolaris
#Phytocoris sulcatus

black cherry

Lygocoris communis
Lygus lineolaris

choke cherry

#Deraeocoris aphidiphagus
#Deraeocoris brevis
Lygocoris communis
Lygus lineolaris
#Phytocoris canadensis
Taedia scrupea

peach

#Blepharidopterus angulatus
*#*Campylomma verbasci*
#Ceratocapsus digitulus
#Ceratocapsus fuscinus
#Ceratocapsus incisus
#Ceratocapsus pilosulus
#Ceratocapsus pumilus
#Deraeocoris nebulosus
#Diaphnocoris provancheri
#Hyaliodes harti
Lygocoris caryae
Lygocoris communis
Lygocoris omnivagus
Lygocoris quercalbae
Lygus hesperus

Lygus lineolaris
Lygus plagiatus
Lygus shulli
#Phytocoris conspurcatus*
#Phytocoris interspersus*
#Phytocoris lasiomerus*
#Phytocoris neglectus*
#Phytocoris sulcatus*
#Plagiognathus politus
Stenotus binotatus

pin cherry

#Deraeocoris borealis*
#Deraeocoris fasciolus*
Lygocoris communis
Lygus lineolaris

plum

#Atractotomus mali
#Blepharidopterus angulatus*
#Campyloneura virgula*
#Ceratocapsus digitulus*
#Ceratocapsus fuscinus*
#Ceratocapsus modestus*
#Ceratocapsus pilosulus*
#Diaphnocoris provancheri*
#Hyaliodes harti*
#Hyaliodes vitripennis*
Lygocoris communis
Lygocoris viburni
Lygus hesperus
Lygus lineolaris
#Phytocoris dimidiatus*
#Pilophorus confusus*
#Pilophorus perplexus*
#Plagiognathus obscurus
Stenotus binotatus

sour cherry

#Blepharidopterus angulatus*
#Deraeocoris nebulosus*
#Diaphnocoris provancheri*
#Hyaliodes harti*
Lygocoris communis
Lygocoris omnivagus
Lygus lineolaris

sweet cherry

#Blepharidopterus angulatus
#Diaphnocoris provancheri
#Hyaliodes harti
 **Lygocoris communis*
 **Lygocoris omnivagus*
 **Lygus hesperus*
 **Lygus shulli*
 **Lygus lineolaris*
#Phytocoris conspurcatus
#Phytocoris interspersus
#Phytocoris lasiomerus
#Phytocoris neglectus
#Phytocoris sulcatus
 **Stenotus binotatus*

Pyrus spp. — pear

**#Atractotomus mali*
 #Blepharidopterus angulatus
**#Campylomma verbasci*
 #Ceratocapsus digitulus
 #Ceratocapsus fuscinus
 #Ceratocapsus incisus
 #Ceratocapsus modestus
 #Ceratocapsus pilosulus
 #Ceratocapsus pumilus
 #Deraeocoris aphidiphagus
 #Deraeocoris bakeri
 #Deraeocoris brevis
 #Deraeocoris fasciolus
 #Deraeocoris nebulosus
 #Deraeocoris nitenatus
 #Diaphnocoris provancheri
 #Eurychilopterella luridula
 #Hyaliodes harti
 #Hyaliodes vitripennis
 **Lygocoris communis*
 **Lygocoris omnivagus*
 **Lygus hesperus*
 **Lygus lineolaris*
 **Lygus plagiatus*
 **Lygus shulli*
 #Orthotylus nassatus
 #Orthotylus viridinervis
 #Paraproba capitata
 #Phytocoris conspurcatus
 #Phytocoris corticevivens

#Phytocoris cortitectus
#Phytocoris dimidiatus
#Phytocoris husseyi
#Phytocoris lasiomerus
#Phytocoris neglectus
#Phytocoris onustus
#Phytocoris salicis
#Pilophorus perplexus
**#Plagiognathus politus*
**Stenotus binotatus*

Ribes spp.

currant

**#Dicyphus rubi*
**Lopidea dakota*
**Lygocoris communis*
**Lygocoris belfragii*
**Lygus hesperus*
**Lygus lineolaris*
#Phytocoris canadensis
#Plagiognathus ribesi
**Poecilocapsus lineatus*

gooseberry

#Dicyphus discrepans
**Lygocoris belfragii*
**Lygocoris communis*
**Lygus lineolaris*
#Phytocoris canadensis
#Plagiognathus ribesi

Rubus spp.

blackberry

#Dicyphus hesperus
**Lygocoris communis*
**Lygus hesperus*
**Lygus lineolaris*
**Lygus shulli*
**#Plagiognathus chrysanthemi*
**#Plagiognathus obscurus*

loganberry

**#Campylomma verbasci*
**Lygocoris communis*

190

Lygus hesperus
Lygus lineolaris
Lygus shulli
#Plagiognathus chrysanthemi
Plagiognathus obscurus

raspberry

#Ceratocapsus pumilus
#Deraeocoris brevis
#Deraeocoris fasciolus
#Dicyphus discrepans
*#Dicyphus famelicus
#Dicyphus hesperus
*#Dicyphus rubi
#Heterotoma meriopterum
#Hyaliodes harti
*Lopidea dakota
*Lygocoris belfragii
*Lygocoris communis
*Lygus hesperus
*Lygus lineolaris
#Macrolophus tenuicornis
*#Plagiognathus alboradialis
*#Plagiognathus chrysanthemi
*#Plagiognathus obscurus
*#Plagiognathus politus
*Poecilocapsus lineatus
#Psallus salicellus
#Rhinocapsus vanduzeei

thimbleberry

#Crytopeltis bakeri
#Dicyphus discrepans
#Dicyphus hesperus
#Dicyphus rubi
*Lygus lineolaris
*Lygus shulli
*#Plagiognathus chrysanthemi
*Plagiognathus obscurus

Sambucus sp. — elderberry

*Lygocoris communis
*Lygus lineolaris
*Lygus nubilus

***Vaccinium* spp.**

blueberry

*Lygus lineolaris
*Plagiognathus obscurus

cranberry

*Lygus lineolaris
#Rhinocapsus vanduzeei

***Viburnum* sp. — viburnum**

*#Dicyphus rubi
 *Lygocoris belfragii
 *Lygocoris communis
 *Lygocoris knighti
 *Lygocoris omnivagus
 *Lygus lineolaris

***Vitis* spp.**

cultivated grape

 *Campylomma verbasci
#Ceratocapsus incisus
#Ceratocapsus modestus
#Hyaliodes vitripennis
#Hyaliodes harti
 *Lygocoris communis
 *Lygocoris inconspicuus
 *Lygus lineolaris
 *Taedia scrupea

wild grape

#Ceratocapsus fuscinus
#Ceratocapsus incisus
#Ceratocapsus pumilus
 *Lygocoris communis
 *Lygocoris inconspicuus
 *Lygus lineolaris
 *Poecilocapsus lineatus
 *Prepops rubellicollis
 *Taedia scrupea

References

Arneson, A. P., King, K. M., Glen, R., and Paul, L. C. 1939. Insects of the season 1938 in Saskatchewan. Can. Agric. Insect Pest Rev. 17:63–73.

Atkinson, E. T. 1890. Catalogue of the Insecta. II. Order Rhynchota, suborder Hemiptera-Heteroptera, Family Capstidae. J. Asiat. Soc. Beng. 58:25–199.

Blatchley, W. S. 1926. Heteroptera of eastern North America. Nature Publishing Co., Indianapolis, Ind. 1116 pp.

Braimah, S. A., Kelton, L. A., and Stewart, K. R. 1982. The plant bugs on apple trees in Quebec (Heteroptera: Miridae). Naturaliste can.

Brittain, W. H. 1915a. Some Hemiptera attacking the apple. Proc. Nova Scotia ent. Soc. 1:7–47.

Brittain, W. H. 1915b. The green apple bug (*Lygus invitus* Say) in Nova Scotia. A. Rep. ent. Soc. Ont. (1916) 46:65–78.

Buckell, E. R. 1939. Insects of the season 1938 in British Columbia. Can. Agric. Insect Pest Rev. 17:82–86.

Caesar, L. 1912. Some new or unrecorded Ontario insect pests. A. Rep. ent. Soc. Ont. (1913) 43:100–105.

Caesar, L. 1920. Notes on leaf bugs (Miridae) attacking fruit trees. A. Rep. ent. Soc. Ont. (1921) 51:14–16.

Carvalho, J. C. M. 1952. Neotropical Miridae, 50. On the present generic assignment of the species in the Biologica Centrali Americana (Hemiptera). Bolm. Mus. nac. Zool. 118:1–17.

Carvalho, J. C. M. 1957–1959. A catalogue of the Miridae of the world. Parts I–IV. Archos. Mus. nac., Rio de J., Brazil.

Douglas, J. W., and Scott, J. 1875. British Hemiptera: additions and corrections. Entomologist's mon. Mag. 12:100–102.

Downes, W. 1927. A preliminary list of the Heteroptera and Homoptera of British Columbia. Proc. ent. Soc. Br. Columb. 23:1–22.

Downes, W. 1957. Notes on some Hemiptera which have been introduced into British Columbia. Proc. ent. Soc. Br. Columb. 54:11–17.

Emmons, E. 1854. Natural History of New York. Agriculture, Vol. 5. Insects. Albany. 272 pp.

Fabricius, J. C. 1787. Mantissa Insectorum. . . . Hafniae. 472 pp.

Fabricius, J. C. 1794. Entomologia systematica 4. Hafniae. 472 pp.

Fabricius, J. C. 1798. Supplementum entomologiae systematicae. Hafniae. 572 pp.

Fabricius, J. C. 1803. Systema Rhyngotorum Secundum Ordines, Genera, Species. . . . Brunsvigae. Hafniae. 314 pp.

Fallén, C. F. 1807. Monographia Cimicum Sueciae. Apud. C. G. Proft. 123 pp.

Felt, E. P. 1915. Banded grape bug. Bull. Albany St. Mus. 175:41.

Fieber, F. X. 1861. Die europäischen Hemiptera. Halbflüger (Rhynchota Heteroptera). Wien. 444 pp.

Gibson, A. 1905. Injurious insects of the flower garden. A. Rep. ent. Soc. Ont. (1906) 36:105–122.

Gilliatt, F. C. 1930. Insects of the season 1930 in Nova Scotia. A. Rep. ent. Soc. Ont. (1931) 61:10–13.

Gilliatt, F. C. 1935. Some predators of the European red mite, *Paratetranychus pilosus* C. & F., in Nova Scotia. Can. J. Res. D. 13:19–38.

Glen, R., and King, K. M. 1938. Insects of the season 1937 in Saskatchewan. Can. Agric. Insect Pest Rev. 16:52–57. .

Gmelin, J. F. 1788–1793. Caroli a Linne systema naturae. Tome 1. IV: 2041–2224.

Henry, T. J. 1977. *Orthotylus nassatus*, a European plant bug new to North America (Heteroptera: Miridae). USDA Coop. Plant Pest Rep. 2:605–608.

Henry, T. J., and Wheeler, A. G. 1979. Palearctic Miridae in North America; records of newly discovered and little-known species (Hemiptera-Heteroptera). Proc. ent. Soc. Wash. 81:257–268.

Herrich-Schaeffer, G. A. W. 1836–53. Die Wanzenartigen Insecten. Nürnberg. pp 35–116.

Hussey, R. F. 1954. Some new or little-known Miridae from northeastern United States (Hemiptera). Proc. ent. Soc. Wash. 56:196–202.

Kelton, L. A. 1961. A new Nearctic genus of Miridae, with notes on *Diaphnidea* Uhler 1895 and *Brachynotocoris* Reuter 1880 (Hemiptera). Can. Ent. 93:566–568.

Kelton, L. A. 1965. *Diaphnidea* Uhler and *Diaphnocoris* Kelton in North America (Hemiptera: Miridae). Can. Ent. 97:1025–1030.

Kelton, L. A. 1971*a*. Four new species of *Lygocoris* from Canada (Heteroptera: Miridae). Can. Ent. 103:1107–1110.

Kelton, L. A. 1971*b*. Review of *Lygocoris* species found in Canada and Alaska (Heteroptera: Miridae). Mem. ent. Soc. Canada. No. 83. 87 pp.

Kelton, L. A. 1975. The Lygus bugs (Genus *Lygus* Hahn) in North America (Heteroptera: Miridae). Mem. ent. Soc. Canada. No. 95. 101 pp.

Kelton, L. A. 1980*a*. Descriptions of three new species of Miridae from the Prairie Provinces and a new record of European Phylini in the Nearctic region (Heteroptera). Can. Ent. 112:285–292.

Kelton, L. A. 1980*b*. Description of a new species of *Parthenicus* Reuter, new records of Holarctic Orthotylini in Canada, and new synonomy for *Diaphnocoris pellucida* (Heteroptera: Miridae). Can. Ent. 112:341–344.

Kelton, L. A. 1980*c*. The insects and arachnids of Canada. Part 8. The plant bugs of the Prairie Provinces of Canada (Heteroptera: Miridae). Agric. Can. Publ. 1703. 408 pp.

Kelton, L. A. 1982. Description of a new species of *Plagiognathus* Fieber, and additional records of European *Psallus salicellus* in the Nearctic region (Heteroptera: Miridae). Can. Ent. 114:169–172.

Kelton, L. A. 1982. New records of European *Pilophorus* and *Orthotylus* in Canada (Heteroptera: Miridae). Can. Ent. 114:283–287.

King, K. M., and Glen, R. 1936. Insects of the season 1935 in Saskatchewan. Can. Agric. Insect Pest Rev. 14:63–67.

Kirkaldy, G. W. 1906. Notes on the classification and nomenclature of the Hemipterous superfamily Miroidea. Can. Ent. 38:369–376.

Kirschbaum, C. L. 1855. Rhynchotographische Beitrage (Capsinen). Die Rhynchoten der Gegend von Wiesbaden. Jahrb. Ver. Naturk. Herz. Nassau, 10:161–348.

Knight, H. H. 1915. Observations on the oviposition of certain Capsids. J. econ. Ent. 8:293–298.

Knight, H. H. 1916. Remarks on *Lygus invitus* Say, with descriptions of a new species and a variety of *Lygus* (Hemiptera: Miridae). Can. Ent. 48:345–349.

Knight, H. H. 1917*a*. Records of European Miridae occurring in North America (Hemiptera, Miridae). Can. Ent. 49:248–250.

Knight, H. H. 1917*b*. A revision of the genus *Lygus* as it occurs in America north of Mexico, with biological data on the species from New York. Bull. Cornell Univ. agric. Exp. Stn 391:555–645.

Knight, H. H. 1918. An investigation of the scarring of fruit caused by apple redbugs. Bull. Cornell Univ. agric. Exp. Stn 396:187–208.

Knight, H. H. 1920. New and little-known species of *Phytocoris* from the eastern United States (Heteroptera-Miridae). Bull. Brooklyn ent. Soc. 15:49–66.

Knight, H. H. 1921*a*. Monograph of the North American species of *Deraeocoris* (Hemiptera, Miridae). Rep. Minn. St. Ent. pp. 76–210.

Knight, H. H. 1921*b*. Nearctic records of species of Miridae known heretofore only from the Palaearctic region (Heterop.). Can. Ent. 53:280–288.

Knight, H. H. 1922. Studies on insects affecting the fruit of the apple with particular reference to the characteristics of the resulting scars. Bull. Cornell Univ. agric. Exp. Stn 410:447–498.

Knight, H. H. 1923*a*. A fourth paper on the species of *Lopidea* (Heteroptera, Miridae). Ent. News 34:65–72.

Knight, H. H. 1923*b*. The Miridae (or Capsidae) of Connecticut. *In* W. Britton. The Hemiptera, or sucking insects, of Connecticut. Bull. Conn. St. geol. nat. Hist. Surv. No. 34.

Knight, H. H. 1924. *Atractotomus mali* (Meyer) found in Nova Scotia (Heteroptera, Miridae). Bull. Brooklyn ent. Soc. 19:65.

Knight, H. H. 1926. A key to the North American species of *Macrolophus* with descriptions of two new species (Hemiptera, Miridae). Ent. News 37:313–316.

Knight, H. H. 1930*a*. New species of *Ceratocapsus* (Hemiptera, Miridae). Bull. Brooklyn ent. Soc. 25:187–198.

Knight, H. H. 1930*b*. A new key to *Paracalocoris* with descriptions of eight new species (Hemiptera, Miridae). Ann. ent. Soc. Am. 23:810–827.

Knight, H. H. 1941*a*. New species of *Lygus* from western United States (Hemiptera, Miridae). Iowa St. Coll. J. Sci. 15:269–273.

Knight, H. H. 1941*b*. The plant bugs, or Miridae, of Illinois. Bull. Ill. St. nat. Hist. Surv. No. 22.

Knight, H. H. 1943. Five new species of *Dicyphus* from western North America and one new *Cyrtopeltis* (Hemiptera, Miridae). Pan-Pacif. Ent. 19:53–58.

Knight, H. H. 1944. *Lygus* Hahn; six new species from western North America (Hemiptera, Miridae). Iowa St. Coll. J. Sci. 18:471–477.

Knight, H. H. 1968. Taxonomic review: Miridae of the Nevada test site and the western United States. Brigham Young Univ. Sci. Bull. 9. 264 pp.

Linnaeus, C. 1758. Systema naturae. Laurentii Salvii, Holmiae. 824 pp.

Lochhead, W. 1903. A key to the insects affecting the small fruits. A. Rep. ent. Soc. Ont. (1904) 34:74–79.

Lord, F. T. 1949. The influence of spray programs on the fauna of apple orchards in Nova Scotia. III. Mites and their predators. Can. Ent. 81:202–230.

Lord, F. T. 1971. Laboratory tests to compare the predatory value of six mirid species in each stage of development against the winter eggs of the European red mite, *Panonychus ulmi* (Acari:Tetranychidae). Can. Ent. 103:1663–1669.

Martin, J. E. H. 1977. The insects and arachnids of Canada. Part 1. Collecting, preparing, and preserving insects, mites, and spiders. Agric. Can. Publ. 1643. 182 pp.

McAtee, W. L. 1916. Key to the Nearctic species of *Paracalocoris* (Heteroptera; Miridae). Ann. ent. Soc. Am. 9:366–390.

McMullen, R. D., and Jong, C. 1967. The influence of three insecticides on predation of pear psylla, *Psylla pyricola*. Can. Ent. 99:1292–1297.

MacLellan, C. R. 1972. Codling moth populations under natural, integrated, and chemical control on apple in Nova Scotia (Lepidoptera: Olethreutidae). Can. Ent. 104:1397–1404.

MacNay, C. G. 1962. Fruit pests. Can. Agric. Insect Pest Rev. 40:84–85.

MacPhee, A. W., and Sanford, K. H. 1954. The influence of spray programs on the fauna of apple orchards in Nova Scotia. VII. Effects on some beneficial arthropods. Can. Ent. 86:128–135.

MacPhee, A. W., and Sanford, K. H. 1956. The influence of spray programs on the fauna of apple orchards in Nova Scotia. X. Supplement to VII. Effects on some beneficial arthropods. Can. Ent. 88:631–634.

MacPhee, A. W., and Sanford, K. H. 1961. The influence of spray programs on the fauna of apple orchards in Nova Scotia. XII. Second supplement to VII. Effects on beneficial arthropods. Can. Ent. 93:671–673.

Madsen, B. J., and Morgan, C. V. G. 1975. Mites and insects collected from vineyards in the Okanagan and Similkaneen Valleys, British Columbia. J. ent. Br. Columb. 72:9–14.

Meyer, L. R. 1843. Verzeichniss der in der Schweiz einheimischen Rhynchoten (Hemiptera Linn.). Heft 1, familie Capsini. Solothurn. 115 pp.

Osborn, H. 1892. Catalogue of the Hemiptera of Iowa. Proc. Iowa Acad. Sci. 1:120–123.

Palisot de Beauvois, A. M. F. J. 1805–1821. Insectes recueillis en Afrique et en Amérique. . . . Paris. 267 pp.

Patterson, N. A., and Neary, M. E. 1952. Insects of the season 1951 in Nova Scotia. Can. Agric. Insect Pest Rev. 30:103–105.

Pickett, A. D. 1943. Fruit insects. Can. Agric. Insect Pest Rev. 21:226.

Provancher, L. 1872. Descriptions de plusieurs Hémiptères nouveaux. Naturaliste can. IV:73–79; 103–108.

Provancher, L. 1886–89. Petite faune entomologique du Canada. Les Hémiptères: 3. Quebec. 354 pp.

Reuter, O. M. 1875a. Revisio critica Capsinarum, praecipue Scandinaviae et Fenniae. Akad. Afh. Helsingfors 2:1–190.

Reuter, O. M. 1875b. Capsidae ex America boreali in Museo Holmiensi asservatae, descriptae. Öfvers. K. VetenskAkad. Förh. 32:59–92.

Reuter, O. M. 1878. Hemiptera Gymnocerata Europae. Acta Soc. Sci. fenn. 13:1–188.

Reuter, O. M. 1883. Hemiptera Gymnocerata Europae. Acta Soc. Sci. fenn. 13:313–496.

Reuter, O. M. 1888. Revisio synonymica Heteropterorum palaearcticorum. Acta Soc. Sci. fenn. 15:443–812.

Reuter, O. M. 1909. Bermerkungen über nearktische Capsiden nebst Beschreibung neuer Arten. Acta Soc. Sci. fenn. 36:1–86.

Riley, C. V. 1870. Beneficial insects. The glassy winged soldier-bug *Campyloneura vitripennis* (Say). Third Ann. Rep. on the noxious insects of the St. Missouri, p. 137.

Rivard, I., and Paradis, R. O. 1978. Les ravageurs des cultures fruitieres dans le sud-ouest du Quebec en 1977. Ann. Ent. Soc. Quebec 23:91–93.

Ross, W. A., and Caesar, L. 1919. Insects of the season in Ontario. A. Rep. ent. Soc. Ont. (1920) 50:95–104.

Ross, W. A., and Caesar, L. 1920. Notes on leaf bugs (Miridae) attacking fruit trees. A. Rep. ent. Soc. Ont. (1921) 51:14–16.

Ross, W. A., and Caesar, L. 1921. Insects of the season in Ontario. A Rep. ent. Soc. Ont. (1922) 52:42–50.

Sanford, K. H., and Herbert, H. J. 1966. The influence of spray programs on the fauna of apple orchards in Nova Scotia. XV. Chemical controls for winter moth, *Operophtera brumata* (L.), and their effects on phytophagous mite and predator populations. Can. Ent. 98:991–999.

Say, T. 1832. Descriptions of new species of heteropterous Hemiptera of North America. New Harmony. 39 pp.

Scopoli, J. A. 1763. Entomologia Carniolica Exhibens Insecta Carnioliae, etc. Vindobonae. 420 pp.

Shull, W. E. 1933. An investigation of the *Lygus* species which are pests of beans (Hemiptera, Miridae). Univ. Idaho agric. Exp. Stn Res. Bull. 11:1–42.

Smith, J. B. 1909. A report on the insects of New Jersey. Order Hemiptera. Rep. New Jers. St. Mus., pp. 131–170.

Tonks, N. V. 1952. Annotated list of insects and mites collected on brambles in lower Fraser Valley, British Columbia, 1951. Proc. ent. Soc. Br. Columb. (1953) 49:27–29.

Twinn, C. R. 1938. Fruit insects. Can. Agric. Insect Pest Rev. 16:282–289.

Twinn, C. R. 1939. Fruit insects. Can. Agric. Insect Pest Rev. 17:267–273.

Uhler, P. R. 1872. Notices of the Hemiptera of the western territories of the United States, chiefly from the surveys of Dr. F. V. Hayden. Pages 392–423 *in* F. V. Hayden. Prelim. Rep. U.S. Geol. Surv. Montana.

Uhler, P. R. 1878. Notices of the Hemiptera Heteroptera in the collection of the late T. W. Harris, M.D. Proc. Boston Soc. nat. Hist. 19:365–446.

Uhler, P. R. 1886. Check list of the Hemiptera Heteroptera of North America. Bull. Brooklyn ent. Soc. 32 pp.

Uhler, P. R. 1887. Observations on North American Capsidae, with descriptions of new species. Entomologica am. 3:67–72.

Uhler, P. R. 1890. Observations on North American Capsidae, with descriptions of new species. Trans. Md Acad. Sci. 1:73–88.

Uhler, P. R. 1892. Observations on some remarkable Heteroptera of North America. Trans. Md Acad. Sci. 1:179–184.

Uhler, P. R. 1895. A preliminary list of the Hemiptera of Colorado. Pages 1–137 *in* C. P. Gillette and C. F. Baker. Bull. Colo. St. Univ. agric. Exp. Stn No. 31.

Uhler, P. R. 1904. List of Hemiptera-Heteroptera of Las Vegas Hot Springs, New Mexico, collected by Messrs. E. S. Schwarz and Herbert S. Barber. Proc. U.S. natn. Mus. 27:349–364.

Van Duzee, E. P. 1909. Observations on some Hemiptera taken in Florida in the spring of 1908. Bull. Buffalo Soc. nat. Sci. 9:149–230.

Van Duzee, E. P. 1912. Hemipterological gleanings. Bull. Buffalo Soc. nat. Sci. 10:477–512.

Van Duzee, E. P. 1914. A preliminary list of the Hemiptera of San Diego County, California. Trans. S. Diego Soc. nat. Hist. 2:1–57.

Van Duzee, E. P. 1917. Catalogue of the Hemiptera of America north of Mexico (excepting the Aphididae, Coccidae, and Aleurodidae). Univ. Calif. Publs Ent. Vol. 2.

Van Duzee, E. P. 1920. New hemipterous insects of the genera *Aradus, Phytocoris* and *Camptobrochis*. Proc. Calif. Acad. Sci. 9:331–356.

Wolff, J. F. 1804. Icones Cimicum descriptionibus illustratae. 4:127–166.

Index

(Page numbers of principal entries are in boldface;
synonyms of species are in italic type.)